CONSTRUCTION PROJECT LOG BOOK

BNi Building News

BNi® Building News

EDITOR IN CHIEF
William D. Mahoney, P.E.

TECHNICAL SERVICES
Eric Williams
Rod Yabut

DESIGN
Robert O. Wright

BNi Publications, Inc.
1 800 873 6397

LOS ANGELES
10801 National Blvd., Ste.100
Los Angeles, CA 90064

NEW ENGLAND
PO Box 14527
East Providence, RI 02914

ANAHEIM
1612 S. Clementine St.
Anaheim, CA 92802

WASHINGTON, D.C.
502 Maple Ave. West
Vienna, VA 22180

ISBN 1 55701 425 6

PERSONAL DATA

NAME ______________________________

ADDRESS

Home — Street ______________________________

City ______________________________

State ______________________________ Zip ____________

Business — Street ______________________________

City ______________________________

State ______________________________ Zip ____________

CONTACT NUMBERS

Home ( ) ______________________________

usiness ( ) ______________________________

ax ( ) ______________________________

mail ______________________________

I CASE OF EMERGENCY, PLEASE NOTIFY

ame ______________________________

reet ______________________________

ity ______________________________ State ________ Zip ____________

lephone ( ) ______________________________

CONTENTS

CONTACTS

Name
Address

Phone
Fax
E-mail

Name
Address

Phone
Fax
E-mail

Name
Address

Phone
Fax
E-mail

Name
Address

Phone
Fax
E-mail

Name
Address

Phone
Fax
E-mail

Name
Address

Phone
Fax
E-mail

Name
Address

Phone
Fax
E-mail

Name
Address

Phone
Fax
E-mail

CONTACTS

Name
Address

Phone
Fax
E-mail

Name
Address

Phone
Fax
E-mail

Name
Address

Phone
Fax
E-mail

Name
Address

Phone
Fax
E-mail

Name
Address

Phone
Fax
E-mail

Name
Address

Phone
Fax
E-mail

Name
Address

Phone
Fax
E-mail

Name
Address

Phone
Fax
E-mail

JOB NAME: ______________________ DATE: ______________

CONTRACTOR: ______________________ Weather: AM ______ PM ______

Expenses / Materials

____
____
____
____
____

Material Deliveries

____
____
____
____
____

Equipment Use / Hours

____
____
____
____
____

Equipment Rentals

____
____
____
____
____

Daily Work Force No.

Superintendent	______
Bricklayers	______
Carpenters	______
Masons	______
Electricians	______
Iron Workers	______
Plumbers	______
Others	______
______________	______
Total	______

DAILY WORK LOG

7 AM ______________________________

8 AM ______________________________

9 AM ______________________________

10 AM ______________________________

11 AM ______________________________

12 NOON ______________________________

1 PM ______________________________

2 PM ______________________________

3 PM ______________________________

4 PM ______________________________

5 PM ______________________________

6 PM ______________________________

Delays / Problems ______________________________

Schedule Updates / Progress ______________________________

Extra Work / Authorized by ______________________________

Supervisor's Signature ______________________________

JOB NAME: ____________________ DATE: ____________

CONTRACTOR: ____________________ Weather: AM ______ PM ______

DAILY WORK LOG

Expenses / Materials

Material Deliveries

Equipment Use / Hours

Equipment Rentals

Daily Work Force	No.
Superintendent	
Bricklayers	
Carpenters	
Masons	
Electricians	
Iron Workers	
Plumbers	
Others	
Total	

7 AM

8 AM

9 AM

10 AM

11 AM

12 NOON

1 PM

2 PM

3 PM

4 PM

5 PM

6 PM

Delays / Problems

Schedule Updates / Progress

Extra Work / Authorized by

Supervisor's Signature ____________________

JOB NAME: ______________________ DATE: ______________

CONTRACTOR: ______________________ Weather: AM ______ PM ______

Expenses / Materials

____

Material Deliveries

____

Equipment Use / Hours

____

Equipment Rentals

____

Daily Work Force **No.**

Superintendent	______
Bricklayers	______
Carpenters	______
Masons	______
Electricians	______
Iron Workers	______
Plumbers	______
Others	______
______________	______
Total	______

DAILY WORK LOG

7 AM

8 AM

9 AM

10 AM

11 AM

12 NOON

1 PM

2 PM

3 PM

4 PM

5 PM

6 PM

Delays / Problems

Schedule Updates / Progress

Extra Work / Authorized by

Supervisor's Signature ______________________________

JOB NAME: ______________________ DATE: ______________

CONTRACTOR: ______________________ Weather: AM ______ PM ______

DAILY WORK LOG

Expenses / Materials

Material Deliveries

Equipment Use / Hours

Equipment Rentals

Daily Work Force No.

Superintendent ______

Bricklayers ______

Carpenters ______

Masons ______

Electricians ______

Iron Workers ______

Plumbers ______

Others ______

__________ ______

Total ______

7 AM

8 AM

9 AM

10 AM

11 AM

12 NOON

1 PM

2 PM

3 PM

4 PM

5 PM

6 PM

Delays / Problems

Schedule Updates / Progress

Extra Work / Authorized by

Supervisor's Signature ______________________

JOB NAME: ______________________ DATE: ____________

CONTRACTOR: ______________________ Weather: AM ______ PM ______

DAILY WORK LOG

Expenses / Materials

Material Deliveries

Equipment Use / Hours

Equipment Rentals

Daily Work Force **No.**

Superintendent	______
Bricklayers	______
Carpenters	______
Masons	______
Electricians	______
Iron Workers	______
Plumbers	______
Others	______
______	______
Total	______

7 AM

8 AM

9 AM

10 AM

11 AM

12 NOON

1 PM

2 PM

3 PM

4 PM

5 PM

6 PM

Delays / Problems

Schedule Updates / Progress

Extra Work / Authorized by

Supervisor's Signature ______________________

JOB NAME: ______________________ DATE: ____________

CONTRACTOR: ______________________ Weather: AM ________ PM ________

DAILY WORK LOG

Expenses / Materials

Material Deliveries

Equipment Use / Hours

Equipment Rentals

Daily Work Force	**No.**
Superintendent	____
Bricklayers	____
Carpenters	____
Masons	____
Electricians	____
Iron Workers	____
Plumbers	____
Others	____
________	____
Total	____

7 AM

8 AM

9 AM

10 AM

11 AM

12 NOON

1 PM

2 PM

3 PM

4 PM

5 PM

6 PM

Delays / Problems

Schedule Updates / Progress

Extra Work / Authorized by

Supervisor's Signature ______________________

JOB NAME: ______________________ DATE: ______________

CONTRACTOR: ______________________ Weather: AM ______ PM ______

Expenses / Materials

Material Deliveries

Equipment Use / Hours

Equipment Rentals

Daily Work Force **No.**

Superintendent	
Bricklayers	
Carpenters	
Masons	
Electricians	
Iron Workers	
Plumbers	
Others	
Total	

DAILY WORK LOG

7 AM

8 AM

9 AM

10 AM

11 AM

12 NOON

1 PM

2 PM

3 PM

4 PM

5 PM

6 PM

Delays / Problems

Schedule Updates / Progress

Extra Work / Authorized by

Supervisor's Signature ______________________

JOB NAME: ______________________ DATE: ______________

CONTRACTOR: ______________________ Weather: AM ______ PM ______

DAILY WORK LOG

Expenses / Materials

Material Deliveries

Equipment Use / Hours

Equipment Rentals

Daily Work Force **No.**

Superintendent	____
Bricklayers	____
Carpenters	____
Masons	____
Electricians	____
Iron Workers	____
Plumbers	____
Others	____
________	____
Total	____

7 AM

8 AM

9 AM

10 AM

11 AM

12 NOON

1 PM

2 PM

3 PM

4 PM

5 PM

6 PM

Delays / Problems

Schedule Updates / Progress

Extra Work / Authorized by

Supervisor's Signature ______________________

JOB NAME: ______________________ DATE: ______________

CONTRACTOR: ______________________ Weather: AM ______ PM ______

DAILY WORK LOG

Expenses / Materials

____

____

____

____

____

Material Deliveries

____

____

____

____

____

Equipment Use / Hours

____

____

____

____

____

Equipment Rentals

____

____

____

____

____

Daily Work Force	**No.**
Superintendent	______
Bricklayers	______
Carpenters	______
Masons	______
Electricians	______
Iron Workers	______
Plumbers	______
Others	______
____________	______
Total	______

7 AM ____________________

____

8 AM ____________________

____

9 AM ____________________

____

10 AM ____________________

____

11 AM ____________________

____

12 NOON ____________________

____

1 PM ____________________

____

2 PM ____________________

____

3 PM ____________________

____

4 PM ____________________

____

5 PM ____________________

____

6 PM ____________________

____

Delays / Problems ____________________

____

____

Schedule Updates / Progress ____________________

____

____

Extra Work / Authorized by ____________________

____

Supervisor's Signature ______________________________

JOB NAME: ______________________ DATE: ______________

CONTRACTOR: ______________________ Weather: AM ________ PM ________

DAILY WORK LOG

Expenses / Materials

Material Deliveries

Equipment Use / Hours

Equipment Rentals

Daily Work Force **No.**

Superintendent ____

Bricklayers ____

Carpenters ____

Masons ____

Electricians ____

Iron Workers ____

Plumbers ____

Others ____

__________ ____

Total ____

7 AM

8 AM

9 AM

10 AM

11 AM

12 NOON

1 PM

2 PM

3 PM

4 PM

5 PM

6 PM

Delays / Problems

Schedule Updates / Progress

Extra Work / Authorized by

Supervisor's Signature ______________________

JOB NAME: ______________________ DATE: ____________

CONTRACTOR: ______________________ Weather: AM ______ PM ______

Expenses / Materials

____

____

____

____

____

Material Deliveries

____

____

____

____

____

Equipment Use / Hours

____

____

____

____

____

Equipment Rentals

____

____

____

____

____

Daily Work Force	**No.**
Superintendent	______
Bricklayers	______
Carpenters	______
Masons	______
Electricians	______
Iron Workers	______
Plumbers	______
Others	______
____________	______
Total	______

DAILY WORK LOG

7 AM ______________________

____

8 AM ______________________

____

9 AM ______________________

____

10 AM ______________________

____

11 AM ______________________

____

12 NOON ______________________

____

1 PM ______________________

____

2 PM ______________________

____

3 PM ______________________

____

4 PM ______________________

____

5 PM ______________________

____

6 PM ______________________

____

Delays / Problems ______________________

____

____

Schedule Updates / Progress ______________________

____

____

Extra Work / Authorized by ______________________

____

Supervisor's Signature ______________________

JOB NAME: ______________________ DATE: ______________

CONTRACTOR: ______________________ Weather: AM ________ PM ________

Expenses / Materials

____

____

____

____

Material Deliveries

____

____

____

____

____

Equipment Use / Hours

____

____

____

____

____

Equipment Rentals

____

____

____

____

____

Daily Work Force No.

Superintendent	____
Bricklayers	____
Carpenters	____
Masons	____
Electricians	____
Iron Workers	____
Plumbers	____
Others	____
__________	____
Total	____

DAILY WORK LOG

7 AM ______________________

____

8 AM ______________________

____

9 AM ______________________

____

10 AM ______________________

____

11 AM ______________________

____

12 NOON ______________________

____

1 PM ______________________

____

2 PM ______________________

____

3 PM ______________________

____

4 PM ______________________

____

5 PM ______________________

____

6 PM ______________________

____

Delays / Problems ______________________

____

____

Schedule Updates / Progress ______________________

____

____

Extra Work / Authorized by ______________________

____

Supervisor's Signature ______________________

JOB NAME: ______________________________ DATE: ______________

CONTRACTOR: ______________________________ Weather: AM _______ PM _______

Expenses / Materials

____
____
____
____
____

Material Deliveries

____
____
____
____
____

Equipment Use / Hours

____
____
____
____
____

Equipment Rentals

____
____
____
____
____

Daily Work Force	**No.**
Superintendent	______
Bricklayers	______
Carpenters	______
Masons	______
Electricians	______
Iron Workers	______
Plumbers	______
Others	______
______________	______
Total	______

DAILY WORK LOG

7 AM ______________________________________

8 AM ______________________________________

9 AM ______________________________________

10 AM ______________________________________

11 AM ______________________________________

12 NOON ______________________________________

1 PM ______________________________________

2 PM ______________________________________

3 PM ______________________________________

4 PM ______________________________________

5 PM ______________________________________

6 PM ______________________________________

Delays / Problems

____

Schedule Updates / Progress

____

Extra Work / Authorized by

____

Supervisor's Signature ______________________________________

JOB NAME: ______ DATE: ______

CONTRACTOR: ______ Weather: AM ______ PM ______

DAILY WORK LOG

Expenses / Materials

Material Deliveries

Equipment Use / Hours

Equipment Rentals

Daily Work Force **No.**

Superintendent	______
Bricklayers	______
Carpenters	______
Masons	______
Electricians	______
Iron Workers	______
Plumbers	______
Others	______
______	______
Total	______

7 AM

8 AM

9 AM

10 AM

11 AM

12 NOON

1 PM

2 PM

3 PM

4 PM

5 PM

6 PM

Delays / Problems

Schedule Updates / Progress

Extra Work / Authorized by

Supervisor's Signature ______

JOB NAME: ______________________ DATE: ______________

CONTRACTOR: ______________________ Weather: AM ______ PM ______

DAILY WORK LOG

	Daily Work Log
Expenses / Materials	7 AM
	8 AM
Material Deliveries	9 AM
	10 AM
	11 AM
Equipment Use / Hours	12 NOON
	1 PM
	2 PM
Equipment Rentals	3 PM
	4 PM
	5 PM
	6 PM
	Delays / Problems
	Schedule Updates / Progress
	Extra Work / Authorized by

Daily Work Force	**No.**
Superintendent	______
Bricklayers	______
Carpenters	______
Masons	______
Electricians	______
Iron Workers	______
Plumbers	______
Others	______
______________	______
Total	______

Supervisor's Signature ______________________________

JOB NAME: ______________________ DATE: ______________

CONTRACTOR: ______________________ Weather: AM ________ PM ________

Expenses / Materials

Material Deliveries

Equipment Use / Hours

Equipment Rentals

Daily Work Force No.

Superintendent	______
Bricklayers	______
Carpenters	______
Masons	______
Electricians	______
Iron Workers	______
Plumbers	______
Others	______
______________	______
Total	______

DAILY WORK LOG

7 AM

8 AM

9 AM

10 AM

11 AM

12 NOON

1 PM

2 PM

3 PM

4 PM

5 PM

6 PM

Delays / Problems

Schedule Updates / Progress

Extra Work / Authorized by

Supervisor's Signature ______________________

JOB NAME: ____________ DATE: ____________

CONTRACTOR: ____________ Weather: AM ______ PM ______

DAILY WORK LOG

Expenses / Materials

Material Deliveries

Equipment Use / Hours

Equipment Rentals

Daily Work Force	**No.**
Superintendent	
Bricklayers	
Carpenters	
Masons	
Electricians	
Iron Workers	
Plumbers	
Others	
Total	

7 AM

8 AM

9 AM

10 AM

11 AM

12 NOON

1 PM

2 PM

3 PM

4 PM

5 PM

6 PM

Delays / Problems

Schedule Updates / Progress

Extra Work / Authorized by

Supervisor's Signature ____________

JOB NAME: ______________________ DATE: __________

CONTRACTOR: ______________________ Weather: AM ______ PM ______

Expenses / Materials

Material Deliveries

Equipment Use / Hours

Equipment Rentals

Daily Work Force **No.**

Superintendent	______
Bricklayers	______
Carpenters	______
Masons	______
Electricians	______
Iron Workers	______
Plumbers	______
Others	______
______________	______
Total	______

DAILY WORK LOG

7 AM

8 AM

9 AM

10 AM

11 AM

12 NOON

1 PM

2 PM

3 PM

4 PM

5 PM

6 PM

Delays / Problems

Schedule Updates / Progress

Extra Work / Authorized by

Supervisor's Signature ______________________

JOB NAME: ______________________ DATE: ______________

CONTRACTOR: ______________________ Weather: AM ______ PM ______

DAILY WORK LOG

Expenses / Materials

____

____

____

____

____

Material Deliveries

____

____

____

____

____

Equipment Use / Hours

____

____

____

____

____

Equipment Rentals

____

____

____

____

____

Daily Work Force	**No.**
Superintendent	______
Bricklayers	______
Carpenters	______
Masons	______
Electricians	______
Iron Workers	______
Plumbers	______
Others	______
______________	______
Total	______

7 AM ______________________________

8 AM ______________________________

9 AM ______________________________

10 AM ______________________________

11 AM ______________________________

12 NOON ______________________________

1 PM ______________________________

2 PM ______________________________

3 PM ______________________________

4 PM ______________________________

5 PM ______________________________

6 PM ______________________________

Delays / Problems ______________________________

Schedule Updates / Progress ______________________________

Extra Work / Authorized by ______________________________

Supervisor's Signature ______________________________

JOB NAME: ______________________ DATE: ______________

CONTRACTOR: ______________________ Weather: AM ______ PM ______

DAILY WORK LOG

Expenses / Materials

Material Deliveries

Equipment Use / Hours

Equipment Rentals

Daily Work Force **No.**

Superintendent	____
Bricklayers	____
Carpenters	____
Masons	____
Electricians	____
Iron Workers	____
Plumbers	____
Others	____
________	____
Total	____

7 AM

8 AM

9 AM

10 AM

11 AM

12 NOON

1 PM

2 PM

3 PM

4 PM

5 PM

6 PM

Delays / Problems

Schedule Updates / Progress

Extra Work / Authorized by

Supervisor's Signature ______________________

JOB NAME: ______________________ DATE: ______________

CONTRACTOR: ______________________ Weather: AM ________ PM ________

DAILY WORK LOG

Expenses / Materials

Material Deliveries

Equipment Use / Hours

Equipment Rentals

Daily Work Force	**No.**
Superintendent	
Bricklayers	
Carpenters	
Masons	
Electricians	
Iron Workers	
Plumbers	
Others	
Total	

7 AM

8 AM

9 AM

10 AM

11 AM

12 NOON

1 PM

2 PM

3 PM

4 PM

5 PM

6 PM

Delays / Problems

Schedule Updates / Progress

Extra Work / Authorized by

Supervisor's Signature ______________________

JOB NAME: ______________________ DATE: ______________

CONTRACTOR: ______________________ Weather: AM ______ PM ______

DAILY WORK LOG

Expenses / Materials

Material Deliveries

Equipment Use / Hours

Equipment Rentals

Daily Work Force	**No.**
Superintendent	
Bricklayers	
Carpenters	
Masons	
Electricians	
Iron Workers	
Plumbers	
Others	
Total	

7 AM

8 AM

9 AM

10 AM

11 AM

12 NOON

1 PM

2 PM

3 PM

4 PM

5 PM

6 PM

Delays / Problems

Schedule Updates / Progress

Extra Work / Authorized by

Supervisor's Signature ______________________

JOB NAME: ______________________ DATE: ____________

CONTRACTOR: ______________________ Weather: AM ______ PM ______

Expenses / Materials

____

____

____

____

____

Material Deliveries

____

____

____

____

____

Equipment Use / Hours

____

____

____

____

____

Equipment Rentals

____

____

____

____

____

Daily Work Force	**No.**
Superintendent	______
Bricklayers	______
Carpenters	______
Masons	______
Electricians	______
Iron Workers	______
Plumbers	______
Others	______
____________	______
Total	______

DAILY WORK LOG

7 AM ______________________

____

8 AM ______________________

____

9 AM ______________________

____

10 AM ______________________

____

11 AM ______________________

____

12 NOON ______________________

____

1 PM ______________________

____

2 PM ______________________

____

3 PM ______________________

____

4 PM ______________________

____

5 PM ______________________

____

6 PM ______________________

____

Delays / Problems ______________________

____

____

Schedule Updates / Progress ______________________

____

____

Extra Work / Authorized by ______________________

____

Supervisor's Signature ______________________

JOB NAME: ______________________ DATE: ______________

CONTRACTOR: ______________________ Weather: AM ________ PM ________

DAILY WORK LOG

Expenses / Materials

Material Deliveries

Equipment Use / Hours

Equipment Rentals

Daily Work Force **No.**

Superintendent	____
Bricklayers	____
Carpenters	____
Masons	____
Electricians	____
Iron Workers	____
Plumbers	____
Others	____
________	____
Total	____

7 AM

8 AM

9 AM

10 AM

11 AM

12 NOON

1 PM

2 PM

3 PM

4 PM

5 PM

6 PM

Delays / Problems

Schedule Updates / Progress

Extra Work / Authorized by

Supervisor's Signature ______________________

JOB NAME: ______________________ DATE: ______________

CONTRACTOR: ______________________ Weather: AM ______ PM ______

Expenses / Materials

Material Deliveries

Equipment Use / Hours

Equipment Rentals

Daily Work Force No.

Superintendent	______
Bricklayers	______
Carpenters	______
Masons	______
Electricians	______
Iron Workers	______
Plumbers	______
Others	______
______	______
Total	______

DAILY WORK LOG

7 AM

8 AM

9 AM

10 AM

11 AM

12 NOON

1 PM

2 PM

3 PM

4 PM

5 PM

6 PM

Delays / Problems

Schedule Updates / Progress

Extra Work / Authorized by

Supervisor's Signature ______________________

JOB NAME: ______________________ DATE: ______________

CONTRACTOR: ______________________ Weather: AM ________ PM ________

DAILY WORK LOG

Expenses / Materials

Material Deliveries

Equipment Use / Hours

Equipment Rentals

Daily Work Force **No.**

Superintendent ____

Bricklayers ____

Carpenters ____

Masons ____

Electricians ____

Iron Workers ____

Plumbers ____

Others ____

________ ____

Total ____

7 AM

8 AM

9 AM

10 AM

11 AM

12 NOON

1 PM

2 PM

3 PM

4 PM

5 PM

6 PM

Delays / Problems

Schedule Updates / Progress

Extra Work / Authorized by

Supervisor's Signature ______________________

JOB NAME: ______________________ DATE: ____________

CONTRACTOR: ______________________ Weather: AM ______ PM ______

DAILY WORK LOG

Expenses / Materials

Material Deliveries

Equipment Use / Hours

Equipment Rentals

Daily Work Force	**No.**
Superintendent	____
Bricklayers	____
Carpenters	____
Masons	____
Electricians	____
Iron Workers	____
Plumbers	____
Others	____
________	____
Total	____

7 AM

8 AM

9 AM

10 AM

11 AM

12 NOON

1 PM

2 PM

3 PM

4 PM

5 PM

6 PM

Delays / Problems

Schedule Updates / Progress

Extra Work / Authorized by

Supervisor's Signature ______________________

JOB NAME: ______________________ DATE: __________

CONTRACTOR: ______________________ Weather: AM ______ PM ______

DAILY WORK LOG

Expenses / Materials

Material Deliveries

Equipment Use / Hours

Equipment Rentals

Daily Work Force **No.**

Superintendent	
Bricklayers	
Carpenters	
Masons	
Electricians	
Iron Workers	
Plumbers	
Others	
Total	

7 AM

8 AM

9 AM

10 AM

11 AM

12 NOON

1 PM

2 PM

3 PM

4 PM

5 PM

6 PM

Delays / Problems

Schedule Updates / Progress

Extra Work / Authorized by

Supervisor's Signature ______________________

JOB NAME: ______________________ DATE: ______________

CONTRACTOR: ______________________ Weather: AM ______ PM ______

Expenses / Materials

____
____
____
____
____

Material Deliveries

____
____
____
____
____

Equipment Use / Hours

____
____
____
____
____

Equipment Rentals

____
____
____
____
____

Daily Work Force	**No.**
Superintendent	______
Bricklayers	______
Carpenters	______
Masons	______
Electricians	______
Iron Workers	______
Plumbers	______
Others	______
______________	______
Total	______

DAILY WORK LOG

7 AM ______________________________

8 AM ______________________________

9 AM ______________________________

10 AM ______________________________

11 AM ______________________________

12 NOON ______________________________

1 PM ______________________________

2 PM ______________________________

3 PM ______________________________

4 PM ______________________________

5 PM ______________________________

6 PM ______________________________

Delays / Problems ______________________________

Schedule Updates / Progress ______________________________

Extra Work / Authorized by ______________________________

Supervisor's Signature ______________________________

JOB NAME: ______________________ DATE: ______________

CONTRACTOR: ______________________ Weather: AM ________ PM ________

DAILY WORK LOG

Expenses / Materials

7 AM

8 AM

Material Deliveries

9 AM

10 AM

11 AM

Equipment Use / Hours

12 NOON

1 PM

2 PM

Equipment Rentals

3 PM

4 PM

5 PM

Daily Work Force No.

Superintendent ______

Bricklayers ______

Carpenters ______

Masons ______

Electricians ______

Iron Workers ______

Plumbers ______

Others ______

______ ______

Total ______

6 PM

Delays / Problems

Schedule Updates / Progress

Extra Work / Authorized by

Supervisor's Signature ______________________

JOB NAME: ______________________ DATE: ______________

CONTRACTOR: ______________________ Weather: AM ______ PM ______

Expenses / Materials

______

______

______

______

______

Material Deliveries

______

______

______

______

______

Equipment Use / Hours

______

______

______

______

______

Equipment Rentals

______

______

______

______

______

Daily Work Force	**No.**
Superintendent	______
Bricklayers	______
Carpenters	______
Masons	______
Electricians	______
Iron Workers	______
Plumbers	______
Others	______
______	______
Total	______

DAILY WORK LOG

7 AM ______

8 AM ______

9 AM ______

10 AM ______

11 AM ______

12 NOON ______

1 PM ______

2 PM ______

3 PM ______

4 PM ______

5 PM ______

6 PM ______

Delays / Problems ______

Schedule Updates / Progress ______

Extra Work / Authorized by ______

Supervisor's Signature ______________________

JOB NAME: ____________________ DATE: ____________

CONTRACTOR: ____________________ Weather: AM ______ PM ______

DAILY WORK LOG

Expenses / Materials

Material Deliveries

Equipment Use / Hours

Equipment Rentals

Daily Work Force **No.**

Superintendent ______

Bricklayers ______

Carpenters ______

Masons ______

Electricians ______

Iron Workers ______

Plumbers ______

Others ______

____________ ______

Total ______

7 AM

8 AM

9 AM

10 AM

11 AM

12 NOON

1 PM

2 PM

3 PM

4 PM

5 PM

6 PM

Delays / Problems

Schedule Updates / Progress

Extra Work / Authorized by

Supervisor's Signature ____________________

JOB NAME: ______________________ DATE: ______________

CONTRACTOR: ______________________ Weather: AM ________ PM ________

Expenses / Materials

____

____

____

____

____

Material Deliveries

____

____

____

____

____

Equipment Use / Hours

____

____

____

____

____

Equipment Rentals

____

____

____

____

____

Daily Work Force	**No.**
Superintendent	______
Bricklayers	______
Carpenters	______
Masons	______
Electricians	______
Iron Workers	______
Plumbers	______
Others	______
______________	______
Total	______

DAILY WORK LOG

7 AM ______________________

8 AM ______________________

9 AM ______________________

10 AM ______________________

11 AM ______________________

12 NOON ______________________

1 PM ______________________

2 PM ______________________

3 PM ______________________

4 PM ______________________

5 PM ______________________

6 PM ______________________

Delays / Problems ______________________

Schedule Updates / Progress ______________________

Extra Work / Authorized by ______________________

Supervisor's Signature ______________________

JOB NAME: ____________________ DATE: ____________

CONTRACTOR: ____________________ Weather: AM ________ PM ________

Expenses / Materials

DAILY WORK LOG

7 AM

8 AM

9 AM

10 AM

11 AM

12 NOON

1 PM

2 PM

3 PM

4 PM

5 PM

6 PM

Material Deliveries

Equipment Use / Hours

Equipment Rentals

Daily Work Force	**No.**
Superintendent	
Bricklayers	
Carpenters	
Masons	
Electricians	
Iron Workers	
Plumbers	
Others	
Total	

Delays / Problems

Schedule Updates / Progress

Extra Work / Authorized by

Supervisor's Signature ____________________

JOB NAME: ______________________________ DATE: ______________

CONTRACTOR: ______________________________ Weather: AM ______ PM ______

DAILY WORK LOG

7 AM

8 AM

9 AM

10 AM

11 AM

12 NOON

1 PM

2 PM

3 PM

4 PM

5 PM

6 PM

Delays / Problems

Schedule Updates / Progress

Extra Work / Authorized by

Expenses / Materials

Material Deliveries

Equipment Use / Hours

Equipment Rentals

Daily Work Force	**No.**
Superintendent	____
Bricklayers	____
Carpenters	____
Masons	____
Electricians	____
Iron Workers	____
Plumbers	____
Others	____
____________	____
Total	____

Supervisor's Signature ______________________________

JOB NAME: ______________________ DATE: ____________

CONTRACTOR: ______________________ Weather: AM ______ PM ______

DAILY WORK LOG

Expenses / Materials

Material Deliveries

Equipment Use / Hours

Equipment Rentals

Daily Work Force **No.**

Superintendent ______

Bricklayers ______

Carpenters ______

Masons ______

Electricians ______

Iron Workers ______

Plumbers ______

Others ______

____________ ______

Total ______

7 AM

8 AM

9 AM

10 AM

11 AM

12 NOON

1 PM

2 PM

3 PM

4 PM

5 PM

6 PM

Delays / Problems

Schedule Updates / Progress

Extra Work / Authorized by

Supervisor's Signature ______________________

JOB NAME: ______________________ DATE: ______________

CONTRACTOR: ______________________ Weather: AM ______ PM ______

DAILY WORK LOG

Expenses / Materials

Material Deliveries

Equipment Use / Hours

Equipment Rentals

Daily Work Force	**No.**
Superintendent	
Bricklayers	
Carpenters	
Masons	
Electricians	
Iron Workers	
Plumbers	
Others	
Total	

7 AM

8 AM

9 AM

10 AM

11 AM

12 NOON

1 PM

2 PM

3 PM

4 PM

5 PM

6 PM

Delays / Problems

Schedule Updates / Progress

Extra Work / Authorized by

Supervisor's Signature ______________________

JOB NAME: ______________________ DATE: ____________

CONTRACTOR: ______________________ Weather: AM ______ PM ______

DAILY WORK LOG

Expenses / Materials

Material Deliveries

Equipment Use / Hours

Equipment Rentals

Daily Work Force **No.**

Superintendent ______

Bricklayers ______

Carpenters ______

Masons ______

Electricians ______

Iron Workers ______

Plumbers ______

Others ______

______ ______

Total ______

7 AM

8 AM

9 AM

10 AM

11 AM

12 NOON

1 PM

2 PM

3 PM

4 PM

5 PM

6 PM

Delays / Problems

Schedule Updates / Progress

Extra Work / Authorized by

Supervisor's Signature ______________________

JOB NAME: ______________________ DATE: ____________

CONTRACTOR: ______________________ Weather: AM ______ PM ______

DAILY WORK LOG

Expenses / Materials

____

____

____

____

____

Material Deliveries

____

____

____

____

____

Equipment Use / Hours

____

____

____

____

____

Equipment Rentals

____

____

____

____

____

Daily Work Force	**No.**
Superintendent	______
Bricklayers	______
Carpenters	______
Masons	______
Electricians	______
Iron Workers	______
Plumbers	______
Others	______
______________	______
Total	______

7 AM ______________________

____

8 AM ______________________

____

9 AM ______________________

____

10 AM ______________________

____

11 AM ______________________

____

12 NOON ______________________

____

1 PM ______________________

____

2 PM ______________________

____

3 PM ______________________

____

4 PM ______________________

____

5 PM ______________________

____

6 PM ______________________

____

Delays / Problems ______________________

____

____

Schedule Updates / Progress ______________________

____

____

Extra Work / Authorized by ______________________

____

Supervisor's Signature ______________________

JOB NAME: ______________________ DATE: ______________

CONTRACTOR: ______________________ Weather: AM ________ PM ________

DAILY WORK LOG

Expenses / Materials

Material Deliveries

Equipment Use / Hours

Equipment Rentals

Daily Work Force	**No.**
Superintendent	
Bricklayers	
Carpenters	
Masons	
Electricians	
Iron Workers	
Plumbers	
Others	
Total	

7 AM

8 AM

9 AM

10 AM

11 AM

12 NOON

1 PM

2 PM

3 PM

4 PM

5 PM

6 PM

Delays / Problems

Schedule Updates / Progress

Extra Work / Authorized by

Supervisor's Signature ______________________

JOB NAME: ______________________ DATE: __________

CONTRACTOR: ______________________ Weather: AM ______ PM ______

Expenses / Materials

Material Deliveries

Equipment Use / Hours

Equipment Rentals

Daily Work Force **No.**

Superintendent	______
Bricklayers	______
Carpenters	______
Masons	______
Electricians	______
Iron Workers	______
Plumbers	______
Others	______
__________	______
Total	______

DAILY WORK LOG

7 AM

8 AM

9 AM

10 AM

11 AM

12 NOON

1 PM

2 PM

3 PM

4 PM

5 PM

6 PM

Delays / Problems

Schedule Updates / Progress

Extra Work / Authorized by

Supervisor's Signature ______________________

JOB NAME: ______________________ DATE: ______________

CONTRACTOR: ______________________ Weather: AM ________ PM ________

DAILY WORK LOG

Expenses / Materials

Material Deliveries

Equipment Use / Hours

Equipment Rentals

Daily Work Force **No.**

Superintendent ______

Bricklayers ______

Carpenters ______

Masons ______

Electricians ______

Iron Workers ______

Plumbers ______

Others ______

______________ ______

Total ______

7 AM

8 AM

9 AM

10 AM

11 AM

12 NOON

1 PM

2 PM

3 PM

4 PM

5 PM

6 PM

Delays / Problems

Schedule Updates / Progress

Extra Work / Authorized by

Supervisor's Signature ______________________________

JOB NAME: ______________________ DATE: ______________

CONTRACTOR: ______________________ Weather: AM ______ PM ______

Expenses / Materials

Material Deliveries

Equipment Use / Hours

Equipment Rentals

Daily Work Force **No.**

Daily Work Force	No.
Superintendent	
Bricklayers	
Carpenters	
Masons	
Electricians	
Iron Workers	
Plumbers	
Others	
Total	

DAILY WORK LOG

7 AM

8 AM

9 AM

10 AM

11 AM

12 NOON

1 PM

2 PM

3 PM

4 PM

5 PM

6 PM

Delays / Problems

Schedule Updates / Progress

Extra Work / Authorized by

Supervisor's Signature ______________________

JOB NAME: ______________________ DATE: ______________

CONTRACTOR: ______________________ Weather: AM ______ PM ______

DAILY WORK LOG

Expenses / Materials

Material Deliveries

Equipment Use / Hours

Equipment Rentals

Daily Work Force **No.**

Superintendent	____
Bricklayers	____
Carpenters	____
Masons	____
Electricians	____
Iron Workers	____
Plumbers	____
Others	____
________	____
Total	____

7 AM

8 AM

9 AM

10 AM

11 AM

12 NOON

1 PM

2 PM

3 PM

4 PM

5 PM

6 PM

Delays / Problems

Schedule Updates / Progress

Extra Work / Authorized by

Supervisor's Signature ______________________

JOB NAME: ______________________ DATE: ______________

CONTRACTOR: ______________________ Weather: AM ______ PM ______

DAILY WORK LOG

Expenses / Materials

____

____

____

____

____

Material Deliveries

____

____

____

____

____

Equipment Use / Hours

____

____

____

____

____

Equipment Rentals

____

____

____

____

____

Daily Work Force	**No.**
Superintendent	______
Bricklayers	______
Carpenters	______
Masons	______
Electricians	______
Iron Workers	______
Plumbers	______
Others	______
______________	______
Total	______

7 AM ______________________________

8 AM ______________________________

9 AM ______________________________

10 AM ______________________________

11 AM ______________________________

12 NOON ______________________________

1 PM ______________________________

2 PM ______________________________

3 PM ______________________________

4 PM ______________________________

5 PM ______________________________

6 PM ______________________________

Delays / Problems ______________________________

Schedule Updates / Progress ______________________________

Extra Work / Authorized by ______________________________

Supervisor's Signature ______________________________

JOB NAME: ______________________ DATE: __________

CONTRACTOR: ______________________ Weather: AM ______ PM ______

Expenses / Materials

Material Deliveries

Equipment Use / Hours

Equipment Rentals

Daily Work Force **No.**

Superintendent ______

Bricklayers ______

Carpenters ______

Masons ______

Electricians ______

Iron Workers ______

Plumbers ______

Others ______

______ ______

Total ______

DAILY WORK LOG

7 AM

8 AM

9 AM

10 AM

11 AM

12 NOON

1 PM

2 PM

3 PM

4 PM

5 PM

6 PM

Delays / Problems

Schedule Updates / Progress

Extra Work / Authorized by

Supervisor's Signature ______________________

JOB NAME: ______________________________ DATE: ______________

CONTRACTOR: ______________________________ Weather: AM ________ PM ________

DAILY WORK LOG

Expenses / Materials

____

____

____

____

____

Material Deliveries

____

____

____

____

____

Equipment Use / Hours

____

____

____

____

____

Equipment Rentals

____

____

____

____

____

Daily Work Force	**No.**
Superintendent	______
Bricklayers	______
Carpenters	______
Masons	______
Electricians	______
Iron Workers	______
Plumbers	______
Others	______
__________	______
Total	______

7 AM ______________________________

8 AM ______________________________

9 AM ______________________________

10 AM ______________________________

11 AM ______________________________

12 NOON ______________________________

1 PM ______________________________

2 PM ______________________________

3 PM ______________________________

4 PM ______________________________

5 PM ______________________________

6 PM ______________________________

Delays / Problems ______________________________

Schedule Updates / Progress ______________________________

Extra Work / Authorized by ______________________________

Supervisor's Signature ______________________________

JOB NAME: ______________________ DATE: __________

CONTRACTOR: ______________________ Weather: AM ______ PM ______

DAILY WORK LOG

Expenses / Materials

Material Deliveries

Equipment Use / Hours

Equipment Rentals

Daily Work Force **No.**

Superintendent	____
Bricklayers	____
Carpenters	____
Masons	____
Electricians	____
Iron Workers	____
Plumbers	____
Others	____
________	____
Total	____

7 AM

8 AM

9 AM

10 AM

11 AM

12 NOON

1 PM

2 PM

3 PM

4 PM

5 PM

6 PM

Delays / Problems

Schedule Updates / Progress

Extra Work / Authorized by

Supervisor's Signature ______________________

JOB NAME: ______________________ DATE: ______________

CONTRACTOR: ______________________ Weather: AM ________ PM ________

Expenses / Materials

____

____

____

____

____

Material Deliveries

____

____

____

____

____

Equipment Use / Hours

____

____

____

____

____

Equipment Rentals

____

____

____

____

____

Daily Work Force	**No.**
Superintendent	______
Bricklayers	______
Carpenters	______
Masons	______
Electricians	______
Iron Workers	______
Plumbers	______
Others	______
______________	______
Total	______

DAILY WORK LOG

7 AM ______________________________________

____

8 AM ______________________________________

____

9 AM ______________________________________

____

10 AM ______________________________________

____

11 AM ______________________________________

____

12 NOON ______________________________________

____

1 PM ______________________________________

____

2 PM ______________________________________

____

3 PM ______________________________________

____

4 PM ______________________________________

____

5 PM ______________________________________

____

6 PM ______________________________________

____

Delays / Problems ______________________________________

____

____

Schedule Updates / Progress ______________________________________

____

____

Extra Work / Authorized by ______________________________________

____

Supervisor's Signature ______________________________________

JOB NAME: ______________________________ DATE: ______________

CONTRACTOR: ______________________________ Weather: AM ______ PM ______

Expenses / Materials

Material Deliveries

Equipment Use / Hours

Equipment Rentals

Daily Work Force No.

Superintendent ______
Bricklayers ______
Carpenters ______
Masons ______
Electricians ______
Iron Workers ______
Plumbers ______
Others ______
______________ ______

Total ______

DAILY WORK LOG

7 AM

8 AM

9 AM

10 AM

11 AM

12 NOON

1 PM

2 PM

3 PM

4 PM

5 PM

6 PM

Delays / Problems

Schedule Updates / Progress

Extra Work / Authorized by

Supervisor's Signature ______________________________

JOB NAME: ______________________ DATE: ______________

CONTRACTOR: ______________________ Weather: AM ______ PM ______

DAILY WORK LOG

Expenses / Materials

____

____

____

____

____

Material Deliveries

____

____

____

____

____

Equipment Use / Hours

____

____

____

____

____

Equipment Rentals

____

____

____

____

____

Daily Work Force	**No.**
Superintendent	______
Bricklayers	______
Carpenters	______
Masons	______
Electricians	______
Iron Workers	______
Plumbers	______
Others	______
______________	______
Total	______

7 AM ______________________________________

____

8 AM ______________________________________

____

9 AM ______________________________________

____

10 AM ______________________________________

____

11 AM ______________________________________

____

12 NOON ______________________________________

____

1 PM ______________________________________

____

2 PM ______________________________________

____

3 PM ______________________________________

____

4 PM ______________________________________

____

5 PM ______________________________________

____

6 PM ______________________________________

____

Delays / Problems ______________________________________

____

____

Schedule Updates / Progress ______________________________________

____

____

Extra Work / Authorized by ______________________________________

____

Supervisor's Signature ______________________________________

JOB NAME: ______________________ DATE: ______________

CONTRACTOR: ______________________ Weather: AM ________ PM ________

Expenses / Materials

DAILY WORK LOG

7 AM

8 AM

9 AM

Material Deliveries

10 AM

11 AM

Equipment Use / Hours

12 NOON

1 PM

2 PM

Equipment Rentals

3 PM

4 PM

5 PM

Daily Work Force **No.**

6 PM

Superintendent	______
Bricklayers	______
Carpenters	______
Masons	______
Electricians	______
Iron Workers	______
Plumbers	______
Others	______
____________	______
Total	______

Delays / Problems

Schedule Updates / Progress

Extra Work / Authorized by

Supervisor's Signature ______________________________

JOB NAME: ______________________ DATE: ____________

CONTRACTOR: ______________________ Weather: AM ______ PM ______

Expenses / Materials

Material Deliveries

Equipment Use / Hours

Equipment Rentals

Daily Work Force **No.**

Superintendent ______

Bricklayers ______

Carpenters ______

Masons ______

Electricians ______

Iron Workers ______

Plumbers ______

Others ______

______________ ______

Total ______

DAILY WORK LOG

7 AM

8 AM

9 AM

10 AM

11 AM

12 NOON

1 PM

2 PM

3 PM

4 PM

5 PM

6 PM

Delays / Problems

Schedule Updates / Progress

Extra Work / Authorized by

Supervisor's Signature ______________________

JOB NAME: ______________________ DATE: ______________

CONTRACTOR: ______________________ Weather: AM ______ PM ______

DAILY WORK LOG

Expenses / Materials

____

Material Deliveries

____

Equipment Use / Hours

____

Equipment Rentals

____

Daily Work Force	**No.**
Superintendent	______
Bricklayers	______
Carpenters	______
Masons	______
Electricians	______
Iron Workers	______
Plumbers	______
Others	______
____________	______
Total	______

7 AM ______________________

8 AM ______________________

9 AM ______________________

10 AM ______________________

11 AM ______________________

12 NOON ______________________

1 PM ______________________

2 PM ______________________

3 PM ______________________

4 PM ______________________

5 PM ______________________

6 PM ______________________

Delays / Problems ______________________

Schedule Updates / Progress ______________________

Extra Work / Authorized by ______________________

Supervisor's Signature ______________________

JOB NAME: ______________________ DATE: ______________

CONTRACTOR: ______________________ Weather: AM ______ PM ______

DAILY WORK LOG

Expenses / Materials

Material Deliveries

Equipment Use / Hours

Equipment Rentals

Daily Work Force **No.**

Daily Work Force	No.
Superintendent	
Bricklayers	
Carpenters	
Masons	
Electricians	
Iron Workers	
Plumbers	
Others	
Total	

7 AM

8 AM

9 AM

10 AM

11 AM

12 NOON

1 PM

2 PM

3 PM

4 PM

5 PM

6 PM

Delays / Problems

Schedule Updates / Progress

Extra Work / Authorized by

Supervisor's Signature ______________________

JOB NAME: ______________________ DATE: ______________

CONTRACTOR: ______________________ Weather: AM ______ PM ______

DAILY WORK LOG

Expenses / Materials

Material Deliveries

Equipment Use / Hours

Equipment Rentals

Daily Work Force	**No.**
Superintendent	
Bricklayers	
Carpenters	
Masons	
Electricians	
Iron Workers	
Plumbers	
Others	
Total	

7 AM

8 AM

9 AM

10 AM

11 AM

12 NOON

1 PM

2 PM

3 PM

4 PM

5 PM

6 PM

Delays / Problems

Schedule Updates / Progress

Extra Work / Authorized by

Supervisor's Signature ______________________

JOB NAME: ______________________ DATE: ____________

CONTRACTOR: ______________________ Weather: AM ______ PM ______

DAILY WORK LOG

Expenses / Materials

____

____

____

____

____

Material Deliveries

____

____

____

____

____

Equipment Use / Hours

____

____

____

____

____

Equipment Rentals

____

____

____

____

____

Daily Work Force	**No.**
Superintendent	______
Bricklayers	______
Carpenters	______
Masons	______
Electricians	______
Iron Workers	______
Plumbers	______
Others	______
______________	______
Total	______

7 AM ______________________

____

8 AM ______________________

____

9 AM ______________________

____

10 AM ______________________

____

11 AM ______________________

____

12 NOON ______________________

____

1 PM ______________________

____

2 PM ______________________

____

3 PM ______________________

____

4 PM ______________________

____

5 PM ______________________

____

6 PM ______________________

____

Delays / Problems ______________________

____

____

Schedule Updates / Progress ______________________

____

____

Extra Work / Authorized by ______________________

____

Supervisor's Signature ______________________

JOB NAME: ______________________ DATE: ______________

CONTRACTOR: ______________________ Weather: AM ________ PM ________

Expenses / Materials

Material Deliveries

Equipment Use / Hours

Equipment Rentals

Daily Work Force No.

Daily Work Force	**No.**
Superintendent	
Bricklayers	
Carpenters	
Masons	
Electricians	
Iron Workers	
Plumbers	
Others	
Total	

DAILY WORK LOG

7 AM

8 AM

9 AM

10 AM

11 AM

12 NOON

1 PM

2 PM

3 PM

4 PM

5 PM

6 PM

Delays / Problems

Schedule Updates / Progress

Extra Work / Authorized by

Supervisor's Signature ______________________

JOB NAME: ______________________ DATE: ____________

CONTRACTOR: ______________________ Weather: AM ______ PM ______

Expenses / Materials

______

______

______

______

______

Material Deliveries

______

______

______

______

______

Equipment Use / Hours

______

______

______

______

______

Equipment Rentals

______

______

______

______

______

Daily Work Force	**No.**
Superintendent	______
Bricklayers	______
Carpenters	______
Masons	______
Electricians	______
Iron Workers	______
Plumbers	______
Others	______
______	______
Total	______

DAILY WORK LOG

7 AM ______

______

8 AM ______

______

9 AM ______

______

10 AM ______

______

11 AM ______

______

12 NOON ______

______

1 PM ______

______

2 PM ______

______

3 PM ______

______

4 PM ______

______

5 PM ______

______

6 PM ______

______

Delays / Problems ______

______

______

Schedule Updates / Progress ______

______

______

Extra Work / Authorized by ______

______

Supervisor's Signature ______________________

JOB NAME: ______________________ DATE: ______________

CONTRACTOR: ______________________ Weather: AM ________ PM ________

DAILY WORK LOG

Expenses / Materials

Material Deliveries

Equipment Use / Hours

Equipment Rentals

Daily Work Force **No.**

Superintendent	______
Bricklayers	______
Carpenters	______
Masons	______
Electricians	______
Iron Workers	______
Plumbers	______
Others	______
______________	______
Total	______

7 AM

8 AM

9 AM

10 AM

11 AM

12 NOON

1 PM

2 PM

3 PM

4 PM

5 PM

6 PM

Delays / Problems

Schedule Updates / Progress

Extra Work / Authorized by

Supervisor's Signature ______________________________

JOB NAME: ______________________ DATE: ______________

CONTRACTOR: ______________________ Weather: AM ______ PM ______

DAILY WORK LOG

Expenses / Materials

Material Deliveries

Equipment Use / Hours

Equipment Rentals

Daily Work Force	**No.**
Superintendent	
Bricklayers	
Carpenters	
Masons	
Electricians	
Iron Workers	
Plumbers	
Others	
Total	

7 AM

8 AM

9 AM

10 AM

11 AM

12 NOON

1 PM

2 PM

3 PM

4 PM

5 PM

6 PM

Delays / Problems

Schedule Updates / Progress

Extra Work / Authorized by

Supervisor's Signature ______________________

JOB NAME: ____________________ DATE: ____________

CONTRACTOR: ____________________ Weather: AM ______ PM ______

DAILY WORK LOG

Expenses / Materials

Material Deliveries

Equipment Use / Hours

Equipment Rentals

Daily Work Force **No.**

Superintendent ______

Bricklayers ______

Carpenters ______

Masons ______

Electricians ______

Iron Workers ______

Plumbers ______

Others ______

__________ ______

Total ______

7 AM

8 AM

9 AM

10 AM

11 AM

12 NOON

1 PM

2 PM

3 PM

4 PM

5 PM

6 PM

Delays / Problems

Schedule Updates / Progress

Extra Work / Authorized by

Supervisor's Signature ____________________

JOB NAME: ______________________ DATE: ______________

CONTRACTOR: ______________________ Weather: AM ______ PM ______

DAILY WORK LOG

Expenses / Materials

____

____

____

____

____

Material Deliveries

____

____

____

____

____

Equipment Use / Hours

____

____

____

____

____

Equipment Rentals

____

____

____

____

____

Daily Work Force	**No.**
Superintendent	______
Bricklayers	______
Carpenters	______
Masons	______
Electricians	______
Iron Workers	______
Plumbers	______
Others	______
______________	______
Total	______

7 AM ______________________________________

____

8 AM ______________________________________

____

9 AM ______________________________________

____

10 AM ______________________________________

____

11 AM ______________________________________

____

12 NOON ______________________________________

____

1 PM ______________________________________

____

2 PM ______________________________________

____

3 PM ______________________________________

____

4 PM ______________________________________

____

5 PM ______________________________________

____

6 PM ______________________________________

____

Delays / Problems ______________________________________

____

____

Schedule Updates / Progress ______________________________________

____

____

Extra Work / Authorized by ______________________________________

____

Supervisor's Signature ______________________________________

JOB NAME: ______________________ DATE: ______________

CONTRACTOR: ______________________ Weather: AM ______ PM ______

DAILY WORK LOG

Expenses / Materials

Material Deliveries

Equipment Use / Hours

Equipment Rentals

Daily Work Force	**No.**
Superintendent	
Bricklayers	
Carpenters	
Masons	
Electricians	
Iron Workers	
Plumbers	
Others	
Total	

7 AM

8 AM

9 AM

10 AM

11 AM

12 NOON

1 PM

2 PM

3 PM

4 PM

5 PM

6 PM

Delays / Problems

Schedule Updates / Progress

Extra Work / Authorized by

Supervisor's Signature ______________________

JOB NAME: ______________________ DATE: ______________

CONTRACTOR: ______________________ Weather: AM ______ PM ______

Expenses / Materials

____

____

____

____

____

Material Deliveries

____

____

____

____

____

Equipment Use / Hours

____

____

____

____

____

Equipment Rentals

____

____

____

____

____

Daily Work Force	**No.**
Superintendent	______
Bricklayers	______
Carpenters	______
Masons	______
Electricians	______
Iron Workers	______
Plumbers	______
Others	______
______________	______
Total	______

DAILY WORK LOG

7 AM ______________________________

____

8 AM ______________________________

____

9 AM ______________________________

____

10 AM ______________________________

____

11 AM ______________________________

____

12 NOON ______________________________

____

1 PM ______________________________

____

2 PM ______________________________

____

3 PM ______________________________

____

4 PM ______________________________

____

5 PM ______________________________

____

6 PM ______________________________

____

Delays / Problems ______________________________

____

____

Schedule Updates / Progress ______________________________

____

____

Extra Work / Authorized by ______________________________

____

Supervisor's Signature ______________________________

JOB NAME: ______________________ DATE: ______________

CONTRACTOR: ______________________ Weather: AM ________ PM ________

DAILY WORK LOG

Expenses / Materials

Material Deliveries

Equipment Use / Hours

Equipment Rentals

Daily Work Force **No.**

Superintendent ____

Bricklayers ____

Carpenters ____

Masons ____

Electricians ____

Iron Workers ____

Plumbers ____

Others ____

________ ____

Total ____

7 AM

8 AM

9 AM

10 AM

11 AM

12 NOON

1 PM

2 PM

3 PM

4 PM

5 PM

6 PM

Delays / Problems

Schedule Updates / Progress

Extra Work / Authorized by

Supervisor's Signature ______________________

JOB NAME: ______________________ DATE: ______________

CONTRACTOR: ______________________ Weather: AM ______ PM ______

DAILY WORK LOG

Expenses / Materials

Material Deliveries

Equipment Use / Hours

Equipment Rentals

Daily Work Force **No.**

Superintendent	______
Bricklayers	______
Carpenters	______
Masons	______
Electricians	______
Iron Workers	______
Plumbers	______
Others	______
______________	______
Total	______

7 AM

8 AM

9 AM

10 AM

11 AM

12 NOON

1 PM

2 PM

3 PM

4 PM

5 PM

6 PM

Delays / Problems

Schedule Updates / Progress

Extra Work / Authorized by

Supervisor's Signature ______________________________

JOB NAME: ______________________ DATE: ____________

CONTRACTOR: ______________________ Weather: AM ______ PM ______

DAILY WORK LOG

Expenses / Materials

Material Deliveries

Equipment Use / Hours

Equipment Rentals

Daily Work Force **No.**

Superintendent ______

Bricklayers ______

Carpenters ______

Masons ______

Electricians ______

Iron Workers ______

Plumbers ______

Others ______

____________ ______

Total ______

7 AM

8 AM

9 AM

10 AM

11 AM

12 NOON

1 PM

2 PM

3 PM

4 PM

5 PM

6 PM

Delays / Problems

Schedule Updates / Progress

Extra Work / Authorized by

Supervisor's Signature ______________________

JOB NAME: ______________________ DATE: ______________

CONTRACTOR: ______________________ Weather: AM ________ PM ________

DAILY WORK LOG

Expenses / Materials

Material Deliveries

Equipment Use / Hours

Equipment Rentals

Daily Work Force	No.
Superintendent	
Bricklayers	
Carpenters	
Masons	
Electricians	
Iron Workers	
Plumbers	
Others	
Total	

7 AM

8 AM

9 AM

10 AM

11 AM

12 NOON

1 PM

2 PM

3 PM

4 PM

5 PM

6 PM

Delays / Problems

Schedule Updates / Progress

Extra Work / Authorized by

Supervisor's Signature ______________________

JOB NAME: ______________________ DATE: ______________

CONTRACTOR: ______________________ Weather: AM ________ PM ________

DAILY WORK LOG

Expenses / Materials

Material Deliveries

Equipment Use / Hours

Equipment Rentals

Daily Work Force **No.**

Superintendent	
Bricklayers	
Carpenters	
Masons	
Electricians	
Iron Workers	
Plumbers	
Others	
Total	

7 AM

8 AM

9 AM

10 AM

11 AM

12 NOON

1 PM

2 PM

3 PM

4 PM

5 PM

6 PM

Delays / Problems

Schedule Updates / Progress

Extra Work / Authorized by

Supervisor's Signature ______________________

JOB NAME: ______________________ DATE: ______________

CONTRACTOR: ______________________ Weather: AM ______ PM ______

DAILY WORK LOG

Expenses / Materials	
	7 AM
	8 AM
Material Deliveries	9 AM
	10 AM
	11 AM
Equipment Use / Hours	
	12 NOON
	1 PM
	2 PM
Equipment Rentals	
	3 PM
	4 PM
	5 PM
Daily Work Force **No.**	
Superintendent ______	6 PM
Bricklayers ______	
Carpenters ______	Delays / Problems
Masons ______	
Electricians ______	
Iron Workers ______	Schedule Updates / Progress
Plumbers ______	
Others ______	
______ ______	Extra Work / Authorized by
Total ______	

Supervisor's Signature ______________________________

JOB NAME: ______________________ DATE: ______________

CONTRACTOR: ______________________ Weather: AM ________ PM ________

DAILY WORK LOG

Expenses / Materials

Material Deliveries

Equipment Use / Hours

Equipment Rentals

Daily Work Force **No.**

Superintendent ______

Bricklayers ______

Carpenters ______

Masons ______

Electricians ______

Iron Workers ______

Plumbers ______

Others ______

______________ ______

Total ______

7 AM

8 AM

9 AM

10 AM

11 AM

12 NOON

1 PM

2 PM

3 PM

4 PM

5 PM

6 PM

Delays / Problems

Schedule Updates / Progress

Extra Work / Authorized by

Supervisor's Signature ______________________________

JOB NAME: ______________________ DATE: ______________

CONTRACTOR: ______________________ Weather: AM ______ PM ______

Expenses / Materials

____

____

____

____

____

Material Deliveries

____

____

____

____

____

Equipment Use / Hours

____

____

____

____

____

Equipment Rentals

____

____

____

____

____

Daily Work Force	**No.**
Superintendent	______
Bricklayers	______
Carpenters	______
Masons	______
Electricians	______
Iron Workers	______
Plumbers	______
Others	______
____________	______
Total	______

DAILY WORK LOG

7 AM ______________________

____

8 AM ______________________

____

9 AM ______________________

____

10 AM ______________________

____

11 AM ______________________

____

12 NOON ______________________

____

1 PM ______________________

____

2 PM ______________________

____

3 PM ______________________

____

4 PM ______________________

____

5 PM ______________________

____

6 PM ______________________

____

Delays / Problems ______________________

____

____

Schedule Updates / Progress ______________________

____

____

Extra Work / Authorized by ______________________

____

Supervisor's Signature ______________________

JOB NAME: ______________________ DATE: __________

CONTRACTOR: ______________________ Weather: AM ______ PM ______

DAILY WORK LOG

Expenses / Materials	
	7 AM
	8 AM
	9 AM
Material Deliveries	10 AM
	11 AM
Equipment Use / Hours	12 NOON
	1 PM
	2 PM
Equipment Rentals	3 PM
	4 PM
	5 PM

Daily Work Force **No.**

Superintendent ______

Bricklayers ______

Carpenters ______

Masons ______

Electricians ______

Iron Workers ______

Plumbers ______

Others ______

__________ ______

Total ______

6 PM

Delays / Problems

Schedule Updates / Progress

Extra Work / Authorized by

Supervisor's Signature ______________________

JOB NAME: ____________________ DATE: ____________

CONTRACTOR: ____________________ Weather: AM ______ PM ______

DAILY WORK LOG

Expenses / Materials

____

____

____

____

____

Material Deliveries

____

____

____

____

____

Equipment Use / Hours

____

____

____

____

____

Equipment Rentals

____

____

____

____

____

Daily Work Force	**No.**
Superintendent	______
Bricklayers	______
Carpenters	______
Masons	______
Electricians	______
Iron Workers	______
Plumbers	______
Others	______
______________	______
Total	______

7 AM ______________________________

8 AM ______________________________

9 AM ______________________________

10 AM ______________________________

11 AM ______________________________

12 NOON ______________________________

1 PM ______________________________

2 PM ______________________________

3 PM ______________________________

4 PM ______________________________

5 PM ______________________________

6 PM ______________________________

Delays / Problems ______________________________

Schedule Updates / Progress ______________________________

Extra Work / Authorized by ______________________________

Supervisor's Signature ______________________________

JOB NAME: ______________________ DATE: ____________
CONTRACTOR: ______________________ Weather: AM ______ PM ______

DAILY WORK LOG

Expenses / Materials

Material Deliveries

Equipment Use / Hours

Equipment Rentals

Daily Work Force No.

Superintendent ______
Bricklayers ______
Carpenters ______
Masons ______
Electricians ______
Iron Workers ______
Plumbers ______
Others ______
______________ ______

Total ______

7 AM

8 AM

9 AM

10 AM

11 AM

12 NOON

1 PM

2 PM

3 PM

4 PM

5 PM

6 PM

Delays / Problems

Schedule Updates / Progress

Extra Work / Authorized by

Supervisor's Signature ______________________

JOB NAME: ______________________ DATE: ______________

CONTRACTOR: ______________________ Weather: AM ______ PM ______

DAILY WORK LOG

Expenses / Materials

____

____

____

____

____

Material Deliveries

____

____

____

____

____

Equipment Use / Hours

____

____

____

____

____

Equipment Rentals

____

____

____

____

____

Daily Work Force	**No.**
Superintendent	______
Bricklayers	______
Carpenters	______
Masons	______
Electricians	______
Iron Workers	______
Plumbers	______
Others	______
______________	______
Total	______

7 AM ______________________________________

____

8 AM ______________________________________

____

9 AM ______________________________________

____

10 AM ______________________________________

____

11 AM ______________________________________

____

12 NOON ______________________________________

____

1 PM ______________________________________

____

2 PM ______________________________________

____

3 PM ______________________________________

____

4 PM ______________________________________

____

5 PM ______________________________________

____

6 PM ______________________________________

____

Delays / Problems ______________________________________

____

____

Schedule Updates / Progress ______________________________________

____

____

Extra Work / Authorized by ______________________________________

____

Supervisor's Signature ______________________________________

JOB NAME: ______________________ DATE: ____________

CONTRACTOR: ______________________ Weather: AM ________ PM ________

DAILY WORK LOG

Expenses / Materials

Material Deliveries

Equipment Use / Hours

Equipment Rentals

Daily Work Force	**No.**
Superintendent	
Bricklayers	
Carpenters	
Masons	
Electricians	
Iron Workers	
Plumbers	
Others	
Total	

7 AM

8 AM

9 AM

10 AM

11 AM

12 NOON

1 PM

2 PM

3 PM

4 PM

5 PM

6 PM

Delays / Problems

Schedule Updates / Progress

Extra Work / Authorized by

Supervisor's Signature ______________________

JOB NAME: ______________________ DATE: ______________

CONTRACTOR: ______________________ Weather: AM ______ PM ______

DAILY WORK LOG

Expenses / Materials

____
____
____
____
____

Material Deliveries

____
____
____
____
____

Equipment Use / Hours

____
____
____
____
____

Equipment Rentals

____
____
____
____
____

Daily Work Force	**No.**
Superintendent	______
Bricklayers	______
Carpenters	______
Masons	______
Electricians	______
Iron Workers	______
Plumbers	______
Others	______
______________	______
Total	______

7 AM ______________________

____

8 AM ______________________

____

9 AM ______________________

____

10 AM ______________________

____

11 AM ______________________

____

12 NOON ______________________

____

1 PM ______________________

____

2 PM ______________________

____

3 PM ______________________

____

4 PM ______________________

____

5 PM ______________________

____

6 PM ______________________

____

Delays / Problems ______________________

____

____

Schedule Updates / Progress ______________________

____

____

Extra Work / Authorized by ______________________

____

Supervisor's Signature ______________________

JOB NAME: ______________________ DATE: ______________

CONTRACTOR: ______________________ Weather: AM ______ PM ______

DAILY WORK LOG

Expenses / Materials

Material Deliveries

Equipment Use / Hours

Equipment Rentals

Daily Work Force	**No.**
Superintendent	______
Bricklayers	______
Carpenters	______
Masons	______
Electricians	______
Iron Workers	______
Plumbers	______
Others	______
______________	______
Total	______

7 AM

8 AM

9 AM

10 AM

11 AM

12 NOON

1 PM

2 PM

3 PM

4 PM

5 PM

6 PM

Delays / Problems

Schedule Updates / Progress

Extra Work / Authorized by

Supervisor's Signature ______________________

JOB NAME: ______________________ DATE: ______________

CONTRACTOR: ______________________ Weather: AM ______ PM ______

DAILY WORK LOG

Expenses / Materials

____

____

____

____

____

Material Deliveries

____

____

____

____

____

Equipment Use / Hours

____

____

____

____

____

Equipment Rentals

____

____

____

____

____

Daily Work Force	**No.**
Superintendent	______
Bricklayers	______
Carpenters	______
Masons	______
Electricians	______
Iron Workers	______
Plumbers	______
Others	______
______________	______
Total	______

7 AM ______________________________

____

8 AM ______________________________

____

9 AM ______________________________

____

10 AM ______________________________

____

11 AM ______________________________

____

12 NOON ______________________________

____

1 PM ______________________________

____

2 PM ______________________________

____

3 PM ______________________________

____

4 PM ______________________________

____

5 PM ______________________________

____

6 PM ______________________________

____

Delays / Problems ______________________________

____

____

Schedule Updates / Progress ______________________________

____

____

Extra Work / Authorized by ______________________________

____

Supervisor's Signature ______________________________

JOB NAME: ______________________ DATE: __________

CONTRACTOR: ______________________ Weather: AM ______ PM ______

DAILY WORK LOG

Expenses / Materials

Material Deliveries

Equipment Use / Hours

Equipment Rentals

Daily Work Force **No.**

Daily Work Force	No.
Superintendent	
Bricklayers	
Carpenters	
Masons	
Electricians	
Iron Workers	
Plumbers	
Others	
Total	

7 AM

8 AM

9 AM

10 AM

11 AM

12 NOON

1 PM

2 PM

3 PM

4 PM

5 PM

6 PM

Delays / Problems

Schedule Updates / Progress

Extra Work / Authorized by

Supervisor's Signature ______________________

JOB NAME: ______________________________ DATE: ______________

CONTRACTOR: ______________________________ Weather: AM ________ PM ________

DAILY WORK LOG

Expenses / Materials

____

____

____

____

____

Material Deliveries

____

____

____

____

____

Equipment Use / Hours

____

____

____

____

____

Equipment Rentals

____

____

____

____

____

Daily Work Force	**No.**
Superintendent	______
Bricklayers	______
Carpenters	______
Masons	______
Electricians	______
Iron Workers	______
Plumbers	______
Others	______
______________	______
Total	______

7 AM ______________________________

____

8 AM ______________________________

____

9 AM ______________________________

____

10 AM ______________________________

____

11 AM ______________________________

____

12 NOON ______________________________

____

1 PM ______________________________

____

2 PM ______________________________

____

3 PM ______________________________

____

4 PM ______________________________

____

5 PM ______________________________

____

6 PM ______________________________

____

Delays / Problems ______________________________

____

____

Schedule Updates / Progress ______________________________

____

____

Extra Work / Authorized by ______________________________

____

Supervisor's Signature ______________________________

JOB NAME: ______________________ DATE: ______________

CONTRACTOR: ______________________ Weather: AM ________ PM ________

DAILY WORK LOG

Expenses / Materials

Material Deliveries

Equipment Use / Hours

Equipment Rentals

Daily Work Force **No.**

Superintendent	______
Bricklayers	______
Carpenters	______
Masons	______
Electricians	______
Iron Workers	______
Plumbers	______
Others	______
____________	______
Total	______

7 AM

8 AM

9 AM

10 AM

11 AM

12 NOON

1 PM

2 PM

3 PM

4 PM

5 PM

6 PM

Delays / Problems

Schedule Updates / Progress

Extra Work / Authorized by

Supervisor's Signature ______________________________

JOB NAME: ______________________ DATE: ______________

CONTRACTOR: ______________________ Weather: AM ______ PM ______

DAILY WORK LOG

Expenses / Materials

Material Deliveries

Equipment Use / Hours

Equipment Rentals

Daily Work Force **No.**

Superintendent	
Bricklayers	
Carpenters	
Masons	
Electricians	
Iron Workers	
Plumbers	
Others	
Total	

7 AM

8 AM

9 AM

10 AM

11 AM

12 NOON

1 PM

2 PM

3 PM

4 PM

5 PM

6 PM

Delays / Problems

Schedule Updates / Progress

Extra Work / Authorized by

Supervisor's Signature ______________________

JOB NAME: ______________________ DATE: ____________

CONTRACTOR: ______________________ Weather: AM ______ PM ______

DAILY WORK LOG

Expenses / Materials

Material Deliveries

Equipment Use / Hours

Equipment Rentals

Daily Work Force **No.**

Superintendent ____

Bricklayers ____

Carpenters ____

Masons ____

Electricians ____

Iron Workers ____

Plumbers ____

Others ____

__________ ____

Total ____

7 AM

8 AM

9 AM

10 AM

11 AM

12 NOON

1 PM

2 PM

3 PM

4 PM

5 PM

6 PM

Delays / Problems

Schedule Updates / Progress

Extra Work / Authorized by

Supervisor's Signature ______________________

JOB NAME: ______________________ DATE: ____________

CONTRACTOR: ______________________ Weather: AM ______ PM ______

DAILY WORK LOG

Expenses / Materials

____

____

____

____

____

Material Deliveries

____

____

____

____

____

Equipment Use / Hours

____

____

____

____

____

Equipment Rentals

____

____

____

____

____

Daily Work Force	**No.**
Superintendent	______
Bricklayers	______
Carpenters	______
Masons	______
Electricians	______
Iron Workers	______
Plumbers	______
Others	______
______________	______
Total	______

7 AM ______________________________

____

8 AM ______________________________

____

9 AM ______________________________

____

10 AM ______________________________

____

11 AM ______________________________

____

12 NOON ______________________________

____

1 PM ______________________________

____

2 PM ______________________________

____

3 PM ______________________________

____

4 PM ______________________________

____

5 PM ______________________________

____

6 PM ______________________________

____

Delays / Problems ______________________________

____

____

Schedule Updates / Progress ______________________________

____

____

Extra Work / Authorized by ______________________________

____

Supervisor's Signature ______________________________

JOB NAME: ____________________ DATE: ____________

CONTRACTOR: ____________________ Weather: AM ______ PM ______

DAILY WORK LOG

Expenses / Materials

Material Deliveries

Equipment Use / Hours

Equipment Rentals

Daily Work Force **No.**

Superintendent ______

Bricklayers ______

Carpenters ______

Masons ______

Electricians ______

Iron Workers ______

Plumbers ______

Others ______

______ ______

Total ______

7 AM

8 AM

9 AM

10 AM

11 AM

12 NOON

1 PM

2 PM

3 PM

4 PM

5 PM

6 PM

Delays / Problems

Schedule Updates / Progress

Extra Work / Authorized by

Supervisor's Signature ____________________

JOB NAME: ______________________ DATE: ______________

CONTRACTOR: ______________________ Weather: AM ________ PM ________

DAILY WORK LOG

Expenses / Materials

____
____
____
____
____

Material Deliveries

____
____
____
____
____

Equipment Use / Hours

____
____
____
____
____

Equipment Rentals

____
____
____
____
____

Daily Work Force	**No.**
Superintendent	______
Bricklayers	______
Carpenters	______
Masons	______
Electricians	______
Iron Workers	______
Plumbers	______
Others	______
______________	______
Total	______

7 AM ______________________

8 AM ______________________

9 AM ______________________

10 AM ______________________

11 AM ______________________

12 NOON ______________________

1 PM ______________________

2 PM ______________________

3 PM ______________________

4 PM ______________________

5 PM ______________________

6 PM ______________________

Delays / Problems ______________________

Schedule Updates / Progress ______________________

Extra Work / Authorized by ______________________

Supervisor's Signature ______________________

JOB NAME: ______________________ DATE: ______________

CONTRACTOR: ______________________ Weather: AM ________ PM ________

DAILY WORK LOG

Expenses / Materials

Material Deliveries

Equipment Use / Hours

Equipment Rentals

Daily Work Force **No.**

Superintendent ______

Bricklayers ______

Carpenters ______

Masons ______

Electricians ______

Iron Workers ______

Plumbers ______

Others ______

______________ ______

Total ______

7 AM

8 AM

9 AM

10 AM

11 AM

12 NOON

1 PM

2 PM

3 PM

4 PM

5 PM

6 PM

Delays / Problems

Schedule Updates / Progress

Extra Work / Authorized by

Supervisor's Signature ______________________

JOB NAME: ______________________ DATE: ______________

CONTRACTOR: ______________________ Weather: AM ______ PM ______

Expenses / Materials

____

____

____

____

____

Material Deliveries

____

____

____

____

____

Equipment Use / Hours

____

____

____

____

____

Equipment Rentals

____

____

____

____

____

Daily Work Force	**No.**
Superintendent	______
Bricklayers	______
Carpenters	______
Masons	______
Electricians	______
Iron Workers	______
Plumbers	______
Others	______
____________	______
Total	______

DAILY WORK LOG

7 AM ______________________

____

8 AM ______________________

____

9 AM ______________________

____

10 AM ______________________

____

11 AM ______________________

____

12 NOON ______________________

____

1 PM ______________________

____

2 PM ______________________

____

3 PM ______________________

____

4 PM ______________________

____

5 PM ______________________

____

6 PM ______________________

____

Delays / Problems ______________________

____

____

Schedule Updates / Progress ______________________

____

____

Extra Work / Authorized by ______________________

____

Supervisor's Signature ______________________

JOB NAME: ______________________ DATE: ______________

CONTRACTOR: ______________________ Weather: AM ________ PM ________

DAILY WORK LOG

Expenses / Materials

Material Deliveries

Equipment Use / Hours

Equipment Rentals

Daily Work Force **No.**

Superintendent	____
Bricklayers	____
Carpenters	____
Masons	____
Electricians	____
Iron Workers	____
Plumbers	____
Others	____
__________	____
Total	____

7 AM

8 AM

9 AM

10 AM

11 AM

12 NOON

1 PM

2 PM

3 PM

4 PM

5 PM

6 PM

Delays / Problems

Schedule Updates / Progress

Extra Work / Authorized by

Supervisor's Signature ______________________

JOB NAME: ______________________ DATE: __________

CONTRACTOR: ______________________ Weather: AM ______ PM ______

DAILY WORK LOG

Expenses / Materials

____

____

____

____

____

Material Deliveries

____

____

____

____

____

Equipment Use / Hours

____

____

____

____

____

Equipment Rentals

____

____

____

____

____

Daily Work Force	**No.**
Superintendent	______
Bricklayers	______
Carpenters	______
Masons	______
Electricians	______
Iron Workers	______
Plumbers	______
Others	______
______________	______
Total	______

7 AM ______________________

____

8 AM ______________________

____

9 AM ______________________

____

10 AM ______________________

____

11 AM ______________________

____

12 NOON ______________________

____

1 PM ______________________

____

2 PM ______________________

____

3 PM ______________________

____

4 PM ______________________

____

5 PM ______________________

____

6 PM ______________________

____

Delays / Problems ______________________

____

____

Schedule Updates / Progress ______________________

____

____

Extra Work / Authorized by ______________________

____

Supervisor's Signature ______________________

JOB NAME: ______________________ DATE: ____________

CONTRACTOR: ______________________ Weather: AM ________ PM ________

DAILY WORK LOG

Expenses / Materials

Material Deliveries

Equipment Use / Hours

Equipment Rentals

Daily Work Force **No.**

Superintendent ____

Bricklayers ____

Carpenters ____

Masons ____

Electricians ____

Iron Workers ____

Plumbers ____

Others ____

____________ ____

Total ____

7 AM

8 AM

9 AM

10 AM

11 AM

12 NOON

1 PM

2 PM

3 PM

4 PM

5 PM

6 PM

Delays / Problems

Schedule Updates / Progress

Extra Work / Authorized by

Supervisor's Signature ______________________

JOB NAME: ______________________ DATE: ______________

CONTRACTOR: ______________________ Weather: AM ______ PM ______

DAILY WORK LOG

Expenses / Materials

____
____
____
____
____

Material Deliveries

____
____
____
____
____

Equipment Use / Hours

____
____
____
____
____

Equipment Rentals

____
____
____
____
____

Daily Work Force	**No.**
Superintendent	______
Bricklayers	______
Carpenters	______
Masons	______
Electricians	______
Iron Workers	______
Plumbers	______
Others	______
______________	______
Total	______

7 AM ______________________

____

8 AM ______________________

____

9 AM ______________________

____

10 AM ______________________

____

11 AM ______________________

____

12 NOON ______________________

____

1 PM ______________________

____

2 PM ______________________

____

3 PM ______________________

____

4 PM ______________________

____

5 PM ______________________

____

6 PM ______________________

____

Delays / Problems ______________________

____

____

Schedule Updates / Progress ______________________

____

____

Extra Work / Authorized by ______________________

____

Supervisor's Signature ______________________

JOB NAME: ______________________ DATE: ____________

CONTRACTOR: ______________________ Weather: AM ______ PM ______

DAILY WORK LOG

Expenses / Materials

Material Deliveries

Equipment Use / Hours

Equipment Rentals

Daily Work Force **No.**

Superintendent ______

Bricklayers ______

Carpenters ______

Masons ______

Electricians ______

Iron Workers ______

Plumbers ______

Others ______

__________ ______

Total ______

7 AM

8 AM

9 AM

10 AM

11 AM

12 NOON

1 PM

2 PM

3 PM

4 PM

5 PM

6 PM

Delays / Problems

Schedule Updates / Progress

Extra Work / Authorized by

Supervisor's Signature ______________________

JOB NAME: ____________________ DATE: ____________

CONTRACTOR: ____________________ Weather: AM ______ PM ______

DAILY WORK LOG

Expenses / Materials

____
____
____
____
____

Material Deliveries

____
____
____
____
____

Equipment Use / Hours

____
____
____
____
____

Equipment Rentals

____
____
____
____
____

Daily Work Force	**No.**
Superintendent	______
Bricklayers	______
Carpenters	______
Masons	______
Electricians	______
Iron Workers	______
Plumbers	______
Others	______
______________	______
Total	______

7 AM ______________________________

8 AM ______________________________

9 AM ______________________________

10 AM ______________________________

11 AM ______________________________

12 NOON ______________________________

1 PM ______________________________

2 PM ______________________________

3 PM ______________________________

4 PM ______________________________

5 PM ______________________________

6 PM ______________________________

Delays / Problems ______________________________

Schedule Updates / Progress ______________________________

Extra Work / Authorized by ______________________________

Supervisor's Signature ______________________________

JOB NAME: ______________________ DATE: ____________

CONTRACTOR: ______________________ Weather: AM ______ PM ______

Expenses / Materials

____

____

____

____

____

Material Deliveries

____

____

____

____

____

Equipment Use / Hours

____

____

____

____

____

Equipment Rentals

____

____

____

____

____

Daily Work Force	**No.**
Superintendent	______
Bricklayers	______
Carpenters	______
Masons	______
Electricians	______
Iron Workers	______
Plumbers	______
Others	______
______________	______
Total	______

DAILY WORK LOG

7 AM ______________________________

8 AM ______________________________

9 AM ______________________________

10 AM ______________________________

11 AM ______________________________

12 NOON ______________________________

1 PM ______________________________

2 PM ______________________________

3 PM ______________________________

4 PM ______________________________

5 PM ______________________________

6 PM ______________________________

Delays / Problems ______________________________

Schedule Updates / Progress ______________________________

Extra Work / Authorized by ______________________________

Supervisor's Signature ______________________________

JOB NAME: ______________________ DATE: ______________

CONTRACTOR: ______________________ Weather: AM ______ PM ______

Expenses / Materials

____
____
____
____
____

Material Deliveries

____
____
____
____
____

Equipment Use / Hours

____
____
____
____
____

Equipment Rentals

____
____
____
____
____

Daily Work Force	**No.**
Superintendent	______
Bricklayers	______
Carpenters	______
Masons	______
Electricians	______
Iron Workers	______
Plumbers	______
Others	______
______________	______
Total	______

DAILY WORK LOG

7 AM ______________________________

8 AM ______________________________

9 AM ______________________________

10 AM ______________________________

11 AM ______________________________

12 NOON ______________________________

1 PM ______________________________

2 PM ______________________________

3 PM ______________________________

4 PM ______________________________

5 PM ______________________________

6 PM ______________________________

Delays / Problems ______________________________

Schedule Updates / Progress ______________________________

Extra Work / Authorized by ______________________________

Supervisor's Signature ______________________________

JOB NAME: ______________________ DATE: ______________

CONTRACTOR: ______________________ Weather: AM ________ PM ________

DAILY WORK LOG

Expenses / Materials

Material Deliveries

Equipment Use / Hours

Equipment Rentals

Daily Work Force **No.**

Superintendent	______
Bricklayers	______
Carpenters	______
Masons	______
Electricians	______
Iron Workers	______
Plumbers	______
Others	______
______________	______
Total	______

7 AM

8 AM

9 AM

10 AM

11 AM

12 NOON

1 PM

2 PM

3 PM

4 PM

5 PM

6 PM

Delays / Problems

Schedule Updates / Progress

Extra Work / Authorized by

Supervisor's Signature ______________________________

JOB NAME: ______________________ DATE: ______________

CONTRACTOR: ______________________ Weather: AM ______ PM ______

Expenses / Materials

____

____

____

____

____

Material Deliveries

____

____

____

____

____

Equipment Use / Hours

____

____

____

____

____

Equipment Rentals

____

____

____

____

____

Daily Work Force	**No.**
Superintendent	______
Bricklayers	______
Carpenters	______
Masons	______
Electricians	______
Iron Workers	______
Plumbers	______
Others	______
__________	______
Total	______

DAILY WORK LOG

7 AM ______________________

8 AM ______________________

9 AM ______________________

10 AM ______________________

11 AM ______________________

12 NOON ______________________

1 PM ______________________

2 PM ______________________

3 PM ______________________

4 PM ______________________

5 PM ______________________

6 PM ______________________

Delays / Problems ______________________

Schedule Updates / Progress ______________________

Extra Work / Authorized by ______________________

Supervisor's Signature ______________________

JOB NAME: ______________________ DATE: ____________

CONTRACTOR: ______________________ Weather: AM ________ PM ________

DAILY WORK LOG

Expenses / Materials

Material Deliveries

Equipment Use / Hours

Equipment Rentals

Daily Work Force	**No.**
Superintendent	
Bricklayers	
Carpenters	
Masons	
Electricians	
Iron Workers	
Plumbers	
Others	
Total	

7 AM

8 AM

9 AM

10 AM

11 AM

12 NOON

1 PM

2 PM

3 PM

4 PM

5 PM

6 PM

Delays / Problems

Schedule Updates / Progress

Extra Work / Authorized by

Supervisor's Signature ______________________

JOB NAME: ______________________________ DATE: ______________

CONTRACTOR: ______________________________ Weather: AM ______ PM ______

DAILY WORK LOG

Expenses / Materials

____
____
____
____
____

Material Deliveries

____
____
____
____
____

Equipment Use / Hours

____
____
____
____
____

Equipment Rentals

____
____
____
____
____

Daily Work Force	**No.**
Superintendent	______
Bricklayers	______
Carpenters	______
Masons	______
Electricians	______
Iron Workers	______
Plumbers	______
Others	______
______________	______
Total	______

7 AM ______________________________

____

8 AM ______________________________

____

9 AM ______________________________

____

10 AM ______________________________

____

11 AM ______________________________

____

12 NOON ______________________________

____

1 PM ______________________________

____

2 PM ______________________________

____

3 PM ______________________________

____

4 PM ______________________________

____

5 PM ______________________________

____

6 PM ______________________________

____

Delays / Problems ______________________________

____

____

Schedule Updates / Progress ______________________________

____

____

Extra Work / Authorized by ______________________________

____

Supervisor's Signature ______________________________

JOB NAME: ______________________ DATE: ______________

CONTRACTOR: ______________________ Weather: AM ________ PM ________

DAILY WORK LOG

Expenses / Materials

Material Deliveries

Equipment Use / Hours

Equipment Rentals

Daily Work Force **No.**

Superintendent ______

Bricklayers ______

Carpenters ______

Masons ______

Electricians ______

Iron Workers ______

Plumbers ______

Others ______

______________ ______

Total ______

7 AM

8 AM

9 AM

10 AM

11 AM

12 NOON

1 PM

2 PM

3 PM

4 PM

5 PM

6 PM

Delays / Problems

Schedule Updates / Progress

Extra Work / Authorized by

Supervisor's Signature ______________________

JOB NAME: ______________________ DATE: ______________

CONTRACTOR: ______________________ Weather: AM ______ PM ______

Expenses / Materials

____

____

____

____

____

Material Deliveries

____

____

____

____

____

Equipment Use / Hours

____

____

____

____

____

Equipment Rentals

____

____

____

____

____

Daily Work Force	**No.**
Superintendent	______
Bricklayers	______
Carpenters	______
Masons	______
Electricians	______
Iron Workers	______
Plumbers	______
Others	______
______________	______
Total	______

DAILY WORK LOG

7 AM ______________________________

____

8 AM ______________________________

____

9 AM ______________________________

____

10 AM ______________________________

____

11 AM ______________________________

____

12 NOON ______________________________

____

1 PM ______________________________

____

2 PM ______________________________

____

3 PM ______________________________

____

4 PM ______________________________

____

5 PM ______________________________

____

6 PM ______________________________

____

Delays / Problems ______________________________

____

____

Schedule Updates / Progress ______________________________

____

____

Extra Work / Authorized by ______________________________

____

Supervisor's Signature ______________________________

JOB NAME: ____________________ DATE: ____________

CONTRACTOR: ____________________ Weather: AM ______ PM ______

DAILY WORK LOG

Expenses / Materials

Material Deliveries

Equipment Use / Hours

Equipment Rentals

Daily Work Force **No.**

Superintendent	____
Bricklayers	____
Carpenters	____
Masons	____
Electricians	____
Iron Workers	____
Plumbers	____
Others	____
________	____
Total	____

7 AM

8 AM

9 AM

10 AM

11 AM

12 NOON

1 PM

2 PM

3 PM

4 PM

5 PM

6 PM

Delays / Problems

Schedule Updates / Progress

Extra Work / Authorized by

Supervisor's Signature ____________________

JOB NAME: ______________________ DATE: ______________

CONTRACTOR: ______________________ Weather: AM ______ PM ______

DAILY WORK LOG

Expenses / Materials

____

____

____

____

____

Material Deliveries

____

____

____

____

____

Equipment Use / Hours

____

____

____

____

____

Equipment Rentals

____

____

____

____

____

Daily Work Force	**No.**
Superintendent	______
Bricklayers	______
Carpenters	______
Masons	______
Electricians	______
Iron Workers	______
Plumbers	______
Others	______
______________	______
Total	______

7 AM ______________________________

____

8 AM ______________________________

____

9 AM ______________________________

____

10 AM ______________________________

____

11 AM ______________________________

____

12 NOON ______________________________

____

1 PM ______________________________

____

2 PM ______________________________

____

3 PM ______________________________

____

4 PM ______________________________

____

5 PM ______________________________

____

6 PM ______________________________

____

Delays / Problems ______________________________

____

____

Schedule Updates / Progress ______________________________

____

____

Extra Work / Authorized by ______________________________

____

Supervisor's Signature ______________________________

JOB NAME: ______________________ DATE: ____________

CONTRACTOR: ______________________ Weather: AM ______ PM ______

DAILY WORK LOG

Expenses / Materials

Material Deliveries

Equipment Use / Hours

Equipment Rentals

Daily Work Force	No.
Superintendent	____
Bricklayers	____
Carpenters	____
Masons	____
Electricians	____
Iron Workers	____
Plumbers	____
Others	____
____________	____
Total	____

7 AM

8 AM

9 AM

10 AM

11 AM

12 NOON

1 PM

2 PM

3 PM

4 PM

5 PM

6 PM

Delays / Problems

Schedule Updates / Progress

Extra Work / Authorized by

Supervisor's Signature ______________________

JOB NAME: ______________________ DATE: ______________

CONTRACTOR: ______________________ Weather: AM ______ PM ______

Expenses / Materials

Material Deliveries

Equipment Use / Hours

Equipment Rentals

Daily Work Force	**No.**
Superintendent	
Bricklayers	
Carpenters	
Masons	
Electricians	
Iron Workers	
Plumbers	
Others	
Total	

DAILY WORK LOG

7 AM

8 AM

9 AM

10 AM

11 AM

12 NOON

1 PM

2 PM

3 PM

4 PM

5 PM

6 PM

Delays / Problems

Schedule Updates / Progress

Extra Work / Authorized by

Supervisor's Signature ______________________

JOB NAME: ______________________ DATE: ____________

CONTRACTOR: ______________________ Weather: AM ________ PM ________

DAILY WORK LOG

Expenses / Materials

7 AM

8 AM

Material Deliveries

9 AM

10 AM

11 AM

Equipment Use / Hours

12 NOON

1 PM

2 PM

Equipment Rentals

3 PM

4 PM

5 PM

Daily Work Force **No.**

6 PM

Superintendent ______

Bricklayers ______

Carpenters ______

Masons ______

Electricians ______

Iron Workers ______

Plumbers ______

Others ______

______________ ______

Total ______

Delays / Problems

Schedule Updates / Progress

Extra Work / Authorized by

Supervisor's Signature ______________________

JOB NAME: ______________________ DATE: ______________

CONTRACTOR: ______________________ Weather: AM ______ PM ______

DAILY WORK LOG

Expenses / Materials

____

____

____

____

____

Material Deliveries

____

____

____

____

____

Equipment Use / Hours

____

____

____

____

____

Equipment Rentals

____

____

____

____

____

Daily Work Force	**No.**
Superintendent	______
Bricklayers	______
Carpenters	______
Masons	______
Electricians	______
Iron Workers	______
Plumbers	______
Others	______
____________	______
Total	______

7 AM ______________________

____

8 AM ______________________

____

9 AM ______________________

____

10 AM ______________________

____

11 AM ______________________

____

12 NOON ______________________

____

1 PM ______________________

____

2 PM ______________________

____

3 PM ______________________

____

4 PM ______________________

____

5 PM ______________________

____

6 PM ______________________

____

Delays / Problems ______________________

____

____

Schedule Updates / Progress ______________________

____

____

Extra Work / Authorized by ______________________

____

Supervisor's Signature ______________________

JOB NAME: ______________________ DATE: ______________

CONTRACTOR: ______________________ Weather: AM ______ PM ______

DAILY WORK LOG

Expenses / Materials

Material Deliveries

Equipment Use / Hours

Equipment Rentals

Daily Work Force **No.**

Superintendent ____

Bricklayers ____

Carpenters ____

Masons ____

Electricians ____

Iron Workers ____

Plumbers ____

Others ____

________ ____

Total ____

7 AM

8 AM

9 AM

10 AM

11 AM

12 NOON

1 PM

2 PM

3 PM

4 PM

5 PM

6 PM

Delays / Problems

Schedule Updates / Progress

Extra Work / Authorized by

Supervisor's Signature ______________________

JOB NAME: ______________________ DATE: ______________

CONTRACTOR: ______________________ Weather: AM ______ PM ______

Expenses / Materials

Material Deliveries

Equipment Use / Hours

Equipment Rentals

Daily Work Force No.

Superintendent ______

Bricklayers ______

Carpenters ______

Masons ______

Electricians ______

Iron Workers ______

Plumbers ______

Others ______

______________ ______

Total ______

DAILY WORK LOG

7 AM

8 AM

9 AM

10 AM

11 AM

12 NOON

1 PM

2 PM

3 PM

4 PM

5 PM

6 PM

Delays / Problems

Schedule Updates / Progress

Extra Work / Authorized by

Supervisor's Signature ______________________

JOB NAME: ______________________ DATE: ____________

CONTRACTOR: ______________________ Weather: AM ________ PM ________

DAILY WORK LOG

Expenses / Materials

Material Deliveries

Equipment Use / Hours

Equipment Rentals

Daily Work Force	**No.**
Superintendent	
Bricklayers	
Carpenters	
Masons	
Electricians	
Iron Workers	
Plumbers	
Others	
Total	

7 AM

8 AM

9 AM

10 AM

11 AM

12 NOON

1 PM

2 PM

3 PM

4 PM

5 PM

6 PM

Delays / Problems

Schedule Updates / Progress

Extra Work / Authorized by

Supervisor's Signature ______________________

JOB NAME: ______________________ DATE: ____________

CONTRACTOR: ______________________ Weather: AM ______ PM ______

DAILY WORK LOG

Expenses / Materials	
	7 AM
	8 AM
Material Deliveries	9 AM
	10 AM
	11 AM
Equipment Use / Hours	12 NOON
	1 PM
	2 PM
Equipment Rentals	3 PM
	4 PM
	5 PM
Daily Work Force **No.**	6 PM
Superintendent ____	
Bricklayers ____	Delays / Problems
Carpenters ____	
Masons ____	
Electricians ____	Schedule Updates / Progress
Iron Workers ____	
Plumbers ____	
Others ____	Extra Work / Authorized by
________ ____	
Total ____	

Supervisor's Signature ______________________

JOB NAME: ______________________ DATE: ______________

CONTRACTOR: ______________________ Weather: AM ______ PM ______

DAILY WORK LOG

Expenses / Materials

Material Deliveries

Equipment Use / Hours

Equipment Rentals

Daily Work Force **No.**

Superintendent	____
Bricklayers	____
Carpenters	____
Masons	____
Electricians	____
Iron Workers	____
Plumbers	____
Others	____
________	____
Total	____

7 AM

8 AM

9 AM

10 AM

11 AM

12 NOON

1 PM

2 PM

3 PM

4 PM

5 PM

6 PM

Delays / Problems

Schedule Updates / Progress

Extra Work / Authorized by

Supervisor's Signature ______________________

JOB NAME: ______________________ DATE: ______________

CONTRACTOR: ______________________ Weather: AM ______ PM ______

DAILY WORK LOG

Expenses / Materials	

Material Deliveries

____

____

____

____

____

Equipment Use / Hours

____

____

____

____

____

Equipment Rentals

____

____

____

____

____

Daily Work Force	**No.**
Superintendent	______
Bricklayers	______
Carpenters	______
Masons	______
Electricians	______
Iron Workers	______
Plumbers	______
Others	______
______________	______
Total	______

7 AM ______________________

8 AM ______________________

9 AM ______________________

10 AM ______________________

11 AM ______________________

12 NOON ______________________

1 PM ______________________

2 PM ______________________

3 PM ______________________

4 PM ______________________

5 PM ______________________

6 PM ______________________

Delays / Problems ______________________

Schedule Updates / Progress ______________________

Extra Work / Authorized by ______________________

Supervisor's Signature ______________________

JOB NAME: ______ DATE: ______

CONTRACTOR: ______ Weather: AM ______ PM ______

DAILY WORK LOG

Expenses / Materials

Material Deliveries

Equipment Use / Hours

Equipment Rentals

Daily Work Force	**No.**
Superintendent	
Bricklayers	
Carpenters	
Masons	
Electricians	
Iron Workers	
Plumbers	
Others	
Total	

7 AM

8 AM

9 AM

10 AM

11 AM

12 NOON

1 PM

2 PM

3 PM

4 PM

5 PM

6 PM

Delays / Problems

Schedule Updates / Progress

Extra Work / Authorized by

Supervisor's Signature ______

JOB NAME: ____________________ DATE: ____________

CONTRACTOR: ____________________ Weather: AM ______ PM ______

DAILY WORK LOG

Expenses / Materials

Material Deliveries

Equipment Use / Hours

Equipment Rentals

Daily Work Force	**No.**
Superintendent	______
Bricklayers	______
Carpenters	______
Masons	______
Electricians	______
Iron Workers	______
Plumbers	______
Others	______
______	______
Total	______

7 AM

8 AM

9 AM

10 AM

11 AM

12 NOON

1 PM

2 PM

3 PM

4 PM

5 PM

6 PM

Delays / Problems

Schedule Updates / Progress

Extra Work / Authorized by

Supervisor's Signature ____________________

JOB NAME: ______________________ DATE: ______________

CONTRACTOR: ______________________ Weather: AM ______ PM ______

DAILY WORK LOG

Expenses / Materials

Material Deliveries

Equipment Use / Hours

Equipment Rentals

Daily Work Force No.

Superintendent ______

Bricklayers ______

Carpenters ______

Masons ______

Electricians ______

Iron Workers ______

Plumbers ______

Others ______

______________ ______

Total ______

7 AM

8 AM

9 AM

10 AM

11 AM

12 NOON

1 PM

2 PM

3 PM

4 PM

5 PM

6 PM

Delays / Problems

Schedule Updates / Progress

Extra Work / Authorized by

Supervisor's Signature ______________________

JOB NAME: ______________________ DATE: ____________

CONTRACTOR: ______________________ Weather: AM ______ PM ______

DAILY WORK LOG

Expenses / Materials	
	7 AM
	8 AM
Material Deliveries	9 AM
	10 AM
	11 AM
Equipment Use / Hours	12 NOON
	1 PM
	2 PM
Equipment Rentals	3 PM
	4 PM
	5 PM
	6 PM

Expenses / Materials

Material Deliveries

Equipment Use / Hours

Equipment Rentals

Daily Work Force	No.
Superintendent	
Bricklayers	
Carpenters	
Masons	
Electricians	
Iron Workers	
Plumbers	
Others	
Total	

Delays / Problems

Schedule Updates / Progress

Extra Work / Authorized by

Supervisor's Signature ______________________

JOB NAME: ______________________ DATE: ______________

CONTRACTOR: ______________________ Weather: AM ________ PM ________

DAILY WORK LOG

Expenses / Materials

Material Deliveries

Equipment Use / Hours

Equipment Rentals

Daily Work Force **No.**

Superintendent ______

Bricklayers ______

Carpenters ______

Masons ______

Electricians ______

Iron Workers ______

Plumbers ______

Others ______

______________ ______

Total ______

7 AM

8 AM

9 AM

10 AM

11 AM

12 NOON

1 PM

2 PM

3 PM

4 PM

5 PM

6 PM

Delays / Problems

Schedule Updates / Progress

Extra Work / Authorized by

Supervisor's Signature ______________________

JOB NAME: ______________________ DATE: ____________

CONTRACTOR: ______________________ Weather: AM ______ PM ______

Expenses / Materials

____

____

____

____

____

Material Deliveries

____

____

____

____

____

Equipment Use / Hours

____

____

____

____

____

Equipment Rentals

____

____

____

____

____

Daily Work Force	**No.**
Superintendent	______
Bricklayers	______
Carpenters	______
Masons	______
Electricians	______
Iron Workers	______
Plumbers	______
Others	______
____________	______
Total	______

DAILY WORK LOG

7 AM ______________________

____

8 AM ______________________

____

9 AM ______________________

____

10 AM ______________________

____

11 AM ______________________

____

12 NOON ______________________

____

1 PM ______________________

____

2 PM ______________________

____

3 PM ______________________

____

4 PM ______________________

____

5 PM ______________________

____

6 PM ______________________

____

Delays / Problems ______________________

____

____

Schedule Updates / Progress ______________________

____

____

Extra Work / Authorized by ______________________

____

Supervisor's Signature ______________________

JOB NAME: ______________________ DATE: ______________

CONTRACTOR: ______________________ Weather: AM ______ PM ______

DAILY WORK LOG

Expenses / Materials

Material Deliveries

Equipment Use / Hours

Equipment Rentals

Daily Work Force	No.
Superintendent	
Bricklayers	
Carpenters	
Masons	
Electricians	
Iron Workers	
Plumbers	
Others	
Total	

7 AM

8 AM

9 AM

10 AM

11 AM

12 NOON

1 PM

2 PM

3 PM

4 PM

5 PM

6 PM

Delays / Problems

Schedule Updates / Progress

Extra Work / Authorized by

Supervisor's Signature ______________________

JOB NAME: ______________________ DATE: ______________

CONTRACTOR: ______________________ Weather: AM ______ PM ______

DAILY WORK LOG

Expenses / Materials

____

____

____

____

____

Material Deliveries

____

____

____

____

____

Equipment Use / Hours

____

____

____

____

____

Equipment Rentals

____

____

____

____

____

Daily Work Force	**No.**
Superintendent	______
Bricklayers	______
Carpenters	______
Masons	______
Electricians	______
Iron Workers	______
Plumbers	______
Others	______
______________	______
Total	______

7 AM ______________________________

8 AM ______________________________

9 AM ______________________________

10 AM ______________________________

11 AM ______________________________

12 NOON ______________________________

1 PM ______________________________

2 PM ______________________________

3 PM ______________________________

4 PM ______________________________

5 PM ______________________________

6 PM ______________________________

Delays / Problems ______________________________

Schedule Updates / Progress ______________________________

Extra Work / Authorized by ______________________________

Supervisor's Signature ______________________________

JOB NAME: ____________________ DATE: ____________

CONTRACTOR: ____________________ Weather: AM ______ PM ______

DAILY WORK LOG

Expenses / Materials

Material Deliveries

Equipment Use / Hours

Equipment Rentals

Daily Work Force	**No.**
Superintendent	____
Bricklayers	____
Carpenters	____
Masons	____
Electricians	____
Iron Workers	____
Plumbers	____
Others	____
____________	____
Total	____

7 AM

8 AM

9 AM

10 AM

11 AM

12 NOON

1 PM

2 PM

3 PM

4 PM

5 PM

6 PM

Delays / Problems

Schedule Updates / Progress

Extra Work / Authorized by

Supervisor's Signature ______________________________

JOB NAME: ____________________ DATE: ____________

CONTRACTOR: ____________________ Weather: AM ______ PM ______

DAILY WORK LOG

Expenses / Materials

____
____
____
____
____

Material Deliveries

____
____
____
____
____

Equipment Use / Hours

____
____
____
____
____

Equipment Rentals

____
____
____
____
____

Daily Work Force	**No.**
Superintendent	______
Bricklayers	______
Carpenters	______
Masons	______
Electricians	______
Iron Workers	______
Plumbers	______
Others	______
______________	______
Total	______

7 AM ______________________________

8 AM ______________________________

9 AM ______________________________

10 AM ______________________________

11 AM ______________________________

12 NOON ______________________________

1 PM ______________________________

2 PM ______________________________

3 PM ______________________________

4 PM ______________________________

5 PM ______________________________

6 PM ______________________________

Delays / Problems ______________________________

Schedule Updates / Progress ______________________________

Extra Work / Authorized by ______________________________

Supervisor's Signature ______________________________

JOB NAME: ______________________ DATE: ______________

CONTRACTOR: ______________________ Weather: AM ______ PM ______

DAILY WORK LOG

Expenses / Materials

Material Deliveries

Equipment Use / Hours

Equipment Rentals

Daily Work Force **No.**

Superintendent	____
Bricklayers	____
Carpenters	____
Masons	____
Electricians	____
Iron Workers	____
Plumbers	____
Others	____
________	____
Total	____

7 AM

8 AM

9 AM

10 AM

11 AM

12 NOON

1 PM

2 PM

3 PM

4 PM

5 PM

6 PM

Delays / Problems

Schedule Updates / Progress

Extra Work / Authorized by

Supervisor's Signature ______________________

JOB NAME: ________________________________ DATE: ______________

CONTRACTOR: ______________________________ Weather: AM ________ PM ________

DAILY WORK LOG

Expenses / Materials

____

____

____

____

____

Material Deliveries

____

____

____

____

____

Equipment Use / Hours

____

____

____

____

____

Equipment Rentals

____

____

____

____

____

Daily Work Force	**No.**
Superintendent	______
Bricklayers	______
Carpenters	______
Masons	______
Electricians	______
Iron Workers	______
Plumbers	______
Others	______
______________	______
Total	______

7 AM ________________________________

____

8 AM ________________________________

____

9 AM ________________________________

____

10 AM ________________________________

____

11 AM ________________________________

____

12 NOON ________________________________

____

1 PM ________________________________

____

2 PM ________________________________

____

3 PM ________________________________

____

4 PM ________________________________

____

5 PM ________________________________

____

6 PM ________________________________

____

Delays / Problems ________________________________

____

____

Schedule Updates / Progress ________________________________

____

____

Extra Work / Authorized by ________________________________

____

Supervisor's Signature ________________________________

JOB NAME: ______________________ DATE: ____________

CONTRACTOR: ______________________ Weather: AM ________ PM ________

Expenses / Materials

____

Material Deliveries

____

Equipment Use / Hours

____

Equipment Rentals

____

Daily Work Force No.

Superintendent	______
Bricklayers	______
Carpenters	______
Masons	______
Electricians	______
Iron Workers	______
Plumbers	______
Others	______
______________	______
Total	______

DAILY WORK LOG

7 AM

8 AM

9 AM

10 AM

11 AM

12 NOON

1 PM

2 PM

3 PM

4 PM

5 PM

6 PM

Delays / Problems

Schedule Updates / Progress

Extra Work / Authorized by

Supervisor's Signature ______________________

JOB NAME: ______________________ DATE: ______________

CONTRACTOR: ______________________ Weather: AM ______ PM ______

DAILY WORK LOG

Expenses / Materials	

7 AM

8 AM

9 AM

10 AM

11 AM

12 NOON

1 PM

2 PM

3 PM

4 PM

5 PM

6 PM

Expenses / Materials

Material Deliveries

Equipment Use / Hours

Equipment Rentals

Daily Work Force	**No.**
Superintendent	
Bricklayers	
Carpenters	
Masons	
Electricians	
Iron Workers	
Plumbers	
Others	
Total	

Delays / Problems

Schedule Updates / Progress

Extra Work / Authorized by

Supervisor's Signature ______________________

JOB NAME: ______________________ DATE: ______________

CONTRACTOR: ______________________ Weather: AM ______ PM ______

DAILY WORK LOG

Expenses / Materials

Material Deliveries

Equipment Use / Hours

Equipment Rentals

Daily Work Force **No.**

Superintendent	____
Bricklayers	____
Carpenters	____
Masons	____
Electricians	____
Iron Workers	____
Plumbers	____
Others	____
________	____
Total	____

7 AM

8 AM

9 AM

10 AM

11 AM

12 NOON

1 PM

2 PM

3 PM

4 PM

5 PM

6 PM

Delays / Problems

Schedule Updates / Progress

Extra Work / Authorized by

Supervisor's Signature ______________________

JOB NAME: ______________________ DATE: ______________

CONTRACTOR: ______________________ Weather: AM ______ PM ______

DAILY WORK LOG

Expenses / Materials

____

Material Deliveries

____

Equipment Use / Hours

____

Equipment Rentals

____

Daily Work Force **No.**

Superintendent	______
Bricklayers	______
Carpenters	______
Masons	______
Electricians	______
Iron Workers	______
Plumbers	______
Others	______
______________	______
Total	______

7 AM ______________________

8 AM ______________________

9 AM ______________________

10 AM ______________________

11 AM ______________________

12 NOON ______________________

1 PM ______________________

2 PM ______________________

3 PM ______________________

4 PM ______________________

5 PM ______________________

6 PM ______________________

Delays / Problems ______________________

Schedule Updates / Progress ______________________

Extra Work / Authorized by ______________________

Supervisor's Signature ______________________

JOB NAME: ______________________ DATE: ______________

CONTRACTOR: ______________________ Weather: AM ________ PM ________

DAILY WORK LOG

Expenses / Materials	
	7 AM
	8 AM
Material Deliveries	9 AM
	10 AM
	11 AM
Equipment Use / Hours	12 NOON
	1 PM
	2 PM
Equipment Rentals	3 PM
	4 PM
	5 PM
	6 PM

Daily Work Force **No.**

Superintendent	______
Bricklayers	______
Carpenters	______
Masons	______
Electricians	______
Iron Workers	______
Plumbers	______
Others	______
____________	______
Total	______

Delays / Problems

Schedule Updates / Progress

Extra Work / Authorized by

Supervisor's Signature ______________________________

JOB NAME: ______________________ DATE: ______________

CONTRACTOR: ______________________ Weather: AM ______ PM ______

Expenses / Materials

____

____

____

____

____

Material Deliveries

____

____

____

____

____

Equipment Use / Hours

____

____

____

____

____

Equipment Rentals

____

____

____

____

____

Daily Work Force	**No.**
Superintendent	______
Bricklayers	______
Carpenters	______
Masons	______
Electricians	______
Iron Workers	______
Plumbers	______
Others	______
______________	______
Total	______

DAILY WORK LOG

7 AM ______________________________

8 AM ______________________________

9 AM ______________________________

10 AM ______________________________

11 AM ______________________________

12 NOON ______________________________

1 PM ______________________________

2 PM ______________________________

3 PM ______________________________

4 PM ______________________________

5 PM ______________________________

6 PM ______________________________

Delays / Problems ______________________________

Schedule Updates / Progress ______________________________

Extra Work / Authorized by ______________________________

Supervisor's Signature ______________________________

JOB NAME: ______________________ DATE: ______________

CONTRACTOR: ______________________ Weather: AM ______ PM ______

Expenses / Materials

DAILY WORK LOG

7 AM

8 AM

Material Deliveries

9 AM

10 AM

11 AM

Equipment Use / Hours

12 NOON

1 PM

2 PM

Equipment Rentals

3 PM

4 PM

5 PM

Daily Work Force No.

Superintendent	______
Bricklayers	______
Carpenters	______
Masons	______
Electricians	______
Iron Workers	______
Plumbers	______
Others	______
____________	______
Total	______

6 PM

Delays / Problems

Schedule Updates / Progress

Extra Work / Authorized by

Supervisor's Signature ______________________

JOB NAME: ______________________ DATE: ____________

CONTRACTOR: ______________________ Weather: AM ______ PM ______

Expenses / Materials

______

Material Deliveries

______

Equipment Use / Hours

______

Equipment Rentals

______

Daily Work Force **No.**

Daily Work Force	No.
Superintendent	____
Bricklayers	____
Carpenters	____
Masons	____
Electricians	____
Iron Workers	____
Plumbers	____
Others	____
______	____
Total	____

DAILY WORK LOG

7 AM ______

8 AM ______

9 AM ______

10 AM ______

11 AM ______

12 NOON ______

1 PM ______

2 PM ______

3 PM ______

4 PM ______

5 PM ______

6 PM ______

Delays / Problems ______

Schedule Updates / Progress ______

Extra Work / Authorized by ______

Supervisor's Signature ______________________

JOB NAME: ______________________ DATE: ______________

CONTRACTOR: ______________________ Weather: AM ________ PM ________

DAILY WORK LOG

Expenses / Materials

Material Deliveries

Equipment Use / Hours

Equipment Rentals

Daily Work Force	No.
Superintendent	
Bricklayers	
Carpenters	
Masons	
Electricians	
Iron Workers	
Plumbers	
Others	
Total	

7 AM

8 AM

9 AM

10 AM

11 AM

12 NOON

1 PM

2 PM

3 PM

4 PM

5 PM

6 PM

Delays / Problems

Schedule Updates / Progress

Extra Work / Authorized by

Supervisor's Signature ______________________

JOB NAME: ______________________ DATE: ______________

CONTRACTOR: ______________________ Weather: AM ______ PM ______

DAILY WORK LOG

Expenses / Materials	

Material Deliveries

____

____

____

____

____

Equipment Use / Hours

____

____

____

____

____

Equipment Rentals

____

____

____

____

____

Daily Work Force	**No.**
Superintendent	
Bricklayers	
Carpenters	
Masons	
Electricians	
Iron Workers	
Plumbers	
Others	
Total	

7 AM ______________________

8 AM ______________________

9 AM ______________________

10 AM ______________________

11 AM ______________________

12 NOON ______________________

1 PM ______________________

2 PM ______________________

3 PM ______________________

4 PM ______________________

5 PM ______________________

6 PM ______________________

Delays / Problems ______________________

Schedule Updates / Progress ______________________

Extra Work / Authorized by ______________________

Supervisor's Signature ______________________

JOB NAME: ____________________ DATE: ____________

CONTRACTOR: ____________________ Weather: AM ______ PM ______

DAILY WORK LOG

Expenses / Materials

Material Deliveries

Equipment Use / Hours

Equipment Rentals

Daily Work Force **No.**

Superintendent ______

Bricklayers ______

Carpenters ______

Masons ______

Electricians ______

Iron Workers ______

Plumbers ______

Others ______

______ ______

Total ______

7 AM

8 AM

9 AM

10 AM

11 AM

12 NOON

1 PM

2 PM

3 PM

4 PM

5 PM

6 PM

Delays / Problems

Schedule Updates / Progress

Extra Work / Authorized by

Supervisor's Signature ____________________

JOB NAME: ______________________ DATE: ______________

CONTRACTOR: ______________________ Weather: AM _______ PM _______

DAILY WORK LOG

Expenses / Materials

____

Material Deliveries

____

Equipment Use / Hours

____

Equipment Rentals

____

Daily Work Force	**No.**
Superintendent	______
Bricklayers	______
Carpenters	______
Masons	______
Electricians	______
Iron Workers	______
Plumbers	______
Others	______
______________	______
Total	______

7 AM ______________________

8 AM ______________________

9 AM ______________________

10 AM ______________________

11 AM ______________________

12 NOON ______________________

1 PM ______________________

2 PM ______________________

3 PM ______________________

4 PM ______________________

5 PM ______________________

6 PM ______________________

Delays / Problems ______________________

Schedule Updates / Progress ______________________

Extra Work / Authorized by ______________________

Supervisor's Signature ______________________

JOB NAME: ______________________ DATE: __________

CONTRACTOR: ______________________ Weather: AM ______ PM ______

Expenses / Materials

DAILY WORK LOG

7 AM

8 AM

9 AM

10 AM

11 AM

12 NOON

1 PM

2 PM

3 PM

4 PM

5 PM

6 PM

Delays / Problems

Schedule Updates / Progress

Extra Work / Authorized by

Material Deliveries

Equipment Use / Hours

Equipment Rentals

Daily Work Force	**No.**
Superintendent	____
Bricklayers	____
Carpenters	____
Masons	____
Electricians	____
Iron Workers	____
Plumbers	____
Others	____
________	____
Total	____

Supervisor's Signature ______________________

JOB NAME: ______________________________ DATE: ______________

CONTRACTOR: ______________________________ Weather: AM ______ PM ______

Expenses / Materials

____

____

____

____

____

Material Deliveries

____

____

____

____

____

Equipment Use / Hours

____

____

____

____

____

Equipment Rentals

____

____

____

____

____

Daily Work Force	**No.**
Superintendent	______
Bricklayers	______
Carpenters	______
Masons	______
Electricians	______
Iron Workers	______
Plumbers	______
Others	______
______________	______
Total	______

DAILY WORK LOG

7 AM ______________________________

____

8 AM ______________________________

____

9 AM ______________________________

____

10 AM ______________________________

____

11 AM ______________________________

____

12 NOON ______________________________

____

1 PM ______________________________

____

2 PM ______________________________

____

3 PM ______________________________

____

4 PM ______________________________

____

5 PM ______________________________

____

6 PM ______________________________

____

Delays / Problems ______________________________

____

____

Schedule Updates / Progress ______________________________

____

____

Extra Work / Authorized by ______________________________

____

Supervisor's Signature ______________________________

JOB NAME: ______________________ DATE: ____________

CONTRACTOR: ______________________ Weather: AM ______ PM ______

DAILY WORK LOG

Expenses / Materials

Material Deliveries

Equipment Use / Hours

Equipment Rentals

Daily Work Force No.

Superintendent	
Bricklayers	
Carpenters	
Masons	
Electricians	
Iron Workers	
Plumbers	
Others	
Total	

7 AM

8 AM

9 AM

10 AM

11 AM

12 NOON

1 PM

2 PM

3 PM

4 PM

5 PM

6 PM

Delays / Problems

Schedule Updates / Progress

Extra Work / Authorized by

Supervisor's Signature ______________________

JOB NAME: ______________________ DATE: ______________

CONTRACTOR: ______________________ Weather: AM ______ PM ______

DAILY WORK LOG

Expenses / Materials

Material Deliveries

Equipment Use / Hours

Equipment Rentals

Daily Work Force **No.**

Superintendent	______
Bricklayers	______
Carpenters	______
Masons	______
Electricians	______
Iron Workers	______
Plumbers	______
Others	______
__________	______
Total	______

7 AM

8 AM

9 AM

10 AM

11 AM

12 NOON

1 PM

2 PM

3 PM

4 PM

5 PM

6 PM

Delays / Problems

Schedule Updates / Progress

Extra Work / Authorized by

Supervisor's Signature ______________________

JOB NAME: ______________________ DATE: ______________

CONTRACTOR: ______________________ Weather: AM ________ PM ________

DAILY WORK LOG

Expenses / Materials

Material Deliveries

Equipment Use / Hours

Equipment Rentals

Daily Work Force **No.**

	No.
Superintendent	
Bricklayers	
Carpenters	
Masons	
Electricians	
Iron Workers	
Plumbers	
Others	
Total	

7 AM

8 AM

9 AM

10 AM

11 AM

12 NOON

1 PM

2 PM

3 PM

4 PM

5 PM

6 PM

Delays / Problems

Schedule Updates / Progress

Extra Work / Authorized by

Supervisor's Signature ______________________

JOB NAME: ____________________ DATE: ____________

CONTRACTOR: ____________________ Weather: AM ______ PM ______

Expenses / Materials

____

____

____

____

____

Material Deliveries

____

____

____

____

____

Equipment Use / Hours

____

____

____

____

____

Equipment Rentals

____

____

____

____

____

Daily Work Force	**No.**
Superintendent	______
Bricklayers	______
Carpenters	______
Masons	______
Electricians	______
Iron Workers	______
Plumbers	______
Others	______
______________	______
Total	______

DAILY WORK LOG

7 AM ______________________________

____

8 AM ______________________________

____

9 AM ______________________________

____

10 AM ______________________________

____

11 AM ______________________________

____

12 NOON ______________________________

____

1 PM ______________________________

____

2 PM ______________________________

____

3 PM ______________________________

____

4 PM ______________________________

____

5 PM ______________________________

____

6 PM ______________________________

____

Delays / Problems ______________________________

____

____

Schedule Updates / Progress ______________________________

____

____

Extra Work / Authorized by ______________________________

____

Supervisor's Signature ______________________________

JOB NAME: ______________________ DATE: ______________

CONTRACTOR: ______________________ Weather: AM ______ PM ______

DAILY WORK LOG

Expenses / Materials

Material Deliveries

Equipment Use / Hours

Equipment Rentals

Daily Work Force **No.**

	No.
Superintendent	
Bricklayers	
Carpenters	
Masons	
Electricians	
Iron Workers	
Plumbers	
Others	
Total	

7 AM

8 AM

9 AM

10 AM

11 AM

12 NOON

1 PM

2 PM

3 PM

4 PM

5 PM

6 PM

Delays / Problems

Schedule Updates / Progress

Extra Work / Authorized by

Supervisor's Signature ______________________

JOB NAME: ______________________ DATE: ______________

CONTRACTOR: ______________________ Weather: AM ______ PM ______

DAILY WORK LOG

Expenses / Materials

Material Deliveries

Equipment Use / Hours

Equipment Rentals

Daily Work Force No.

Daily Work Force	No.
Superintendent	
Bricklayers	
Carpenters	
Masons	
Electricians	
Iron Workers	
Plumbers	
Others	
Total	

7 AM

8 AM

9 AM

10 AM

11 AM

12 NOON

1 PM

2 PM

3 PM

4 PM

5 PM

6 PM

Delays / Problems

Schedule Updates / Progress

Extra Work / Authorized by

Supervisor's Signature ______________________

JOB NAME: ______________________ DATE: ____________

CONTRACTOR: ______________________ Weather: AM ______ PM ______

Expenses / Materials

Material Deliveries

Equipment Use / Hours

Equipment Rentals

Daily Work Force No.

Superintendent ______
Bricklayers ______
Carpenters ______
Masons ______
Electricians ______
Iron Workers ______
Plumbers ______
Others ______
______ ______

Total ______

DAILY WORK LOG

7 AM

8 AM

9 AM

10 AM

11 AM

12 NOON

1 PM

2 PM

3 PM

4 PM

5 PM

6 PM

Delays / Problems

Schedule Updates / Progress

Extra Work / Authorized by

Supervisor's Signature ______________________

JOB NAME: ____________________ DATE: ____________

CONTRACTOR: ____________________ Weather: AM ______ PM ______

DAILY WORK LOG

Expenses / Materials

Material Deliveries

Equipment Use / Hours

Equipment Rentals

Daily Work Force **No.**

Superintendent ______

Bricklayers ______

Carpenters ______

Masons ______

Electricians ______

Iron Workers ______

Plumbers ______

Others ______

____________ ______

Total ______

7 AM

8 AM

9 AM

10 AM

11 AM

12 NOON

1 PM

2 PM

3 PM

4 PM

5 PM

6 PM

Delays / Problems

Schedule Updates / Progress

Extra Work / Authorized by

Supervisor's Signature ____________________

JOB NAME: ______________________ DATE: ______________

CONTRACTOR: ______________________ Weather: AM ________ PM ________

DAILY WORK LOG

Expenses / Materials

Material Deliveries

Equipment Use / Hours

Equipment Rentals

Daily Work Force No.

Superintendent ______

Bricklayers ______

Carpenters ______

Masons ______

Electricians ______

Iron Workers ______

Plumbers ______

Others ______

____________ ______

Total ______

7 AM

8 AM

9 AM

10 AM

11 AM

12 NOON

1 PM

2 PM

3 PM

4 PM

5 PM

6 PM

Delays / Problems

Schedule Updates / Progress

Extra Work / Authorized by

Supervisor's Signature ______________________________

JOB NAME: ______________________ DATE: ______________

CONTRACTOR: ______________________ Weather: AM ______ PM ______

DAILY WORK LOG

Expenses / Materials	
	7 AM
	8 AM
	9 AM
Material Deliveries	10 AM
	11 AM
Equipment Use / Hours	12 NOON
	1 PM
	2 PM
Equipment Rentals	3 PM
	4 PM
	5 PM

Daily Work Force	No.
Superintendent	
Bricklayers	
Carpenters	
Masons	
Electricians	
Iron Workers	
Plumbers	
Others	
Total	

6 PM

Delays / Problems

Schedule Updates / Progress

Extra Work / Authorized by

Supervisor's Signature ______________________________

JOB NAME: ______________________ DATE: ____________

CONTRACTOR: ______________________ Weather: AM ________ PM ________

Expenses / Materials

Material Deliveries

Equipment Use / Hours

Equipment Rentals

Daily Work Force No.

Superintendent ______

Bricklayers ______

Carpenters ______

Masons ______

Electricians ______

Iron Workers ______

Plumbers ______

Others ______

____________ ______

Total ______

DAILY WORK LOG

7 AM

8 AM

9 AM

10 AM

11 AM

12 NOON

1 PM

2 PM

3 PM

4 PM

5 PM

6 PM

Delays / Problems

Schedule Updates / Progress

Extra Work / Authorized by

Supervisor's Signature ______________________

JOB NAME: ______________________ DATE: ______________

CONTRACTOR: ______________________ Weather: AM ______ PM ______

Expenses / Materials

____

____

____

____

____

Material Deliveries

____

____

____

____

____

Equipment Use / Hours

____

____

____

____

____

Equipment Rentals

____

____

____

____

____

Daily Work Force	**No.**
Superintendent	______
Bricklayers	______
Carpenters	______
Masons	______
Electricians	______
Iron Workers	______
Plumbers	______
Others	______
______________	______
Total	______

DAILY WORK LOG

7 AM ______________________

____

8 AM ______________________

____

9 AM ______________________

____

10 AM ______________________

____

11 AM ______________________

____

12 NOON ______________________

____

1 PM ______________________

____

2 PM ______________________

____

3 PM ______________________

____

4 PM ______________________

____

5 PM ______________________

____

6 PM ______________________

____

Delays / Problems ______________________

____

____

Schedule Updates / Progress ______________________

____

____

Extra Work / Authorized by ______________________

____

Supervisor's Signature ______________________

JOB NAME: ______________________ DATE: ______________

CONTRACTOR: ______________________ Weather: AM ______ PM ______

DAILY WORK LOG

Expenses / Materials

Material Deliveries

Equipment Use / Hours

Equipment Rentals

Daily Work Force **No.**

Superintendent	____
Bricklayers	____
Carpenters	____
Masons	____
Electricians	____
Iron Workers	____
Plumbers	____
Others	____
________	____
Total	____

7 AM

8 AM

9 AM

10 AM

11 AM

12 NOON

1 PM

2 PM

3 PM

4 PM

5 PM

6 PM

Delays / Problems

Schedule Updates / Progress

Extra Work / Authorized by

Supervisor's Signature ______________________

JOB NAME: ______________________ DATE: ______________

CONTRACTOR: ______________________ Weather: AM ______ PM ______

DAILY WORK LOG

Expenses / Materials

____

____

____

____

____

Material Deliveries

____

____

____

____

____

Equipment Use / Hours

____

____

____

____

____

Equipment Rentals

____

____

____

____

____

Daily Work Force	**No.**
Superintendent	______
Bricklayers	______
Carpenters	______
Masons	______
Electricians	______
Iron Workers	______
Plumbers	______
Others	______
______________	______
Total	______

7 AM ______________________________

____

8 AM ______________________________

____

9 AM ______________________________

____

10 AM ______________________________

____

11 AM ______________________________

____

12 NOON ______________________________

____

1 PM ______________________________

____

2 PM ______________________________

____

3 PM ______________________________

____

4 PM ______________________________

____

5 PM ______________________________

____

6 PM ______________________________

____

Delays / Problems ______________________________

____

____

Schedule Updates / Progress ______________________________

____

____

Extra Work / Authorized by ______________________________

____

Supervisor's Signature ______________________________

JOB NAME: ______________________ DATE: __________

CONTRACTOR: ______________________ Weather: AM ______ PM ______

DAILY WORK LOG

Expenses / Materials

Material Deliveries

Equipment Use / Hours

Equipment Rentals

Daily Work Force **No.**

Superintendent ____

Bricklayers ____

Carpenters ____

Masons ____

Electricians ____

Iron Workers ____

Plumbers ____

Others ____

__________ ____

Total ____

7 AM

8 AM

9 AM

10 AM

11 AM

12 NOON

1 PM

2 PM

3 PM

4 PM

5 PM

6 PM

Delays / Problems

Schedule Updates / Progress

Extra Work / Authorized by

Supervisor's Signature ______________________

JOB NAME: ______________________ DATE: ______________

CONTRACTOR: ______________________ Weather: AM ______ PM ______

Expenses / Materials

Material Deliveries

Equipment Use / Hours

Equipment Rentals

Daily Work Force **No.**

Superintendent	____
Bricklayers	____
Carpenters	____
Masons	____
Electricians	____
Iron Workers	____
Plumbers	____
Others	____
________	____
Total	____

DAILY WORK LOG

7 AM

8 AM

9 AM

10 AM

11 AM

12 NOON

1 PM

2 PM

3 PM

4 PM

5 PM

6 PM

Delays / Problems

Schedule Updates / Progress

Extra Work / Authorized by

Supervisor's Signature ______________________

JOB NAME: ______________________ DATE: ______________

CONTRACTOR: ______________________ Weather: AM ________ PM ________

DAILY WORK LOG

Expenses / Materials

Material Deliveries

Equipment Use / Hours

Equipment Rentals

Daily Work Force **No.**

Superintendent	____
Bricklayers	____
Carpenters	____
Masons	____
Electricians	____
Iron Workers	____
Plumbers	____
Others	____
____________	____
Total	____

7 AM

8 AM

9 AM

10 AM

11 AM

12 NOON

1 PM

2 PM

3 PM

4 PM

5 PM

6 PM

Delays / Problems

Schedule Updates / Progress

Extra Work / Authorized by

Supervisor's Signature ______________________

JOB NAME: ______________________ DATE: ______________

CONTRACTOR: ______________________ Weather: AM ________ PM ________

Expenses / Materials

Material Deliveries

Equipment Use / Hours

Equipment Rentals

Daily Work Force	**No.**
Superintendent	____
Bricklayers	____
Carpenters	____
Masons	____
Electricians	____
Iron Workers	____
Plumbers	____
Others	____
________	____
Total	____

DAILY WORK LOG

7 AM

8 AM

9 AM

10 AM

11 AM

12 NOON

1 PM

2 PM

3 PM

4 PM

5 PM

6 PM

Delays / Problems

Schedule Updates / Progress

Extra Work / Authorized by

Supervisor's Signature ______________________

JOB NAME: ______________________ DATE: ______________

CONTRACTOR: ______________________ Weather: AM ______ PM ______

DAILY WORK LOG

Expenses / Materials

____

____

____

____

____

Material Deliveries

____

____

____

____

____

Equipment Use / Hours

____

____

____

____

____

Equipment Rentals

____

____

____

____

____

Daily Work Force	**No.**
Superintendent	______
Bricklayers	______
Carpenters	______
Masons	______
Electricians	______
Iron Workers	______
Plumbers	______
Others	______
______________	______
Total	______

7 AM ______________________

____

8 AM ______________________

____

9 AM ______________________

____

10 AM ______________________

____

11 AM ______________________

____

12 NOON ______________________

____

1 PM ______________________

____

2 PM ______________________

____

3 PM ______________________

____

4 PM ______________________

____

5 PM ______________________

____

6 PM ______________________

____

Delays / Problems

____

____

Schedule Updates / Progress

____

____

Extra Work / Authorized by

____

____

Supervisor's Signature ______________________

JOB NAME: ______________________________ DATE: ______________

CONTRACTOR: ______________________________ Weather: AM ______ PM ______

Expenses / Materials

____
____
____
____
____

Material Deliveries

____
____
____
____
____

Equipment Use / Hours

____
____
____
____
____

Equipment Rentals

____
____
____
____
____

Daily Work Force	**No.**
Superintendent	______
Bricklayers	______
Carpenters	______
Masons	______
Electricians	______
Iron Workers	______
Plumbers	______
Others	______
______________	______
Total	______

DAILY WORK LOG

7 AM ______________________________

____

8 AM ______________________________

____

9 AM ______________________________

____

10 AM ______________________________

____

11 AM ______________________________

____

12 NOON ______________________________

____

1 PM ______________________________

____

2 PM ______________________________

____

3 PM ______________________________

____

4 PM ______________________________

____

5 PM ______________________________

____

6 PM ______________________________

____

Delays / Problems ______________________________

____

____

Schedule Updates / Progress ______________________________

____

____

Extra Work / Authorized by ______________________________

____

Supervisor's Signature ______________________________

JOB NAME: ______________________ DATE: ______________

CONTRACTOR: ______________________ Weather: AM ______ PM ______

DAILY WORK LOG

Expenses / Materials

Material Deliveries

Equipment Use / Hours

Equipment Rentals

Daily Work Force	**No.**
Superintendent	
Bricklayers	
Carpenters	
Masons	
Electricians	
Iron Workers	
Plumbers	
Others	
Total	

7 AM

8 AM

9 AM

10 AM

11 AM

12 NOON

1 PM

2 PM

3 PM

4 PM

5 PM

6 PM

Delays / Problems

Schedule Updates / Progress

Extra Work / Authorized by

Supervisor's Signature ______________________

JOB NAME: ______________________ DATE: ______________

CONTRACTOR: ______________________ Weather: AM ______ PM ______

Expenses / Materials

Material Deliveries

Equipment Use / Hours

Equipment Rentals

Daily Work Force **No.**

Superintendent ______

Bricklayers ______

Carpenters ______

Masons ______

Electricians ______

Iron Workers ______

Plumbers ______

Others ______

______ ______

Total ______

DAILY WORK LOG

7 AM

8 AM

9 AM

10 AM

11 AM

12 NOON

1 PM

2 PM

3 PM

4 PM

5 PM

6 PM

Delays / Problems

Schedule Updates / Progress

Extra Work / Authorized by

Supervisor's Signature ______________________

JOB NAME: ______________________ DATE: ____________

CONTRACTOR: ______________________ Weather: AM ______ PM ______

Expenses / Materials

Material Deliveries

Equipment Use / Hours

Equipment Rentals

Daily Work Force **No.**

Superintendent	____
Bricklayers	____
Carpenters	____
Masons	____
Electricians	____
Iron Workers	____
Plumbers	____
Others	____
________	____
Total	____

DAILY WORK LOG

7 AM

8 AM

9 AM

10 AM

11 AM

12 NOON

1 PM

2 PM

3 PM

4 PM

5 PM

6 PM

Delays / Problems

Schedule Updates / Progress

Extra Work / Authorized by

Supervisor's Signature ______________________

JOB NAME: ______________________ DATE: ______________

CONTRACTOR: ______________________ Weather: AM ______ PM ______

Expenses / Materials

Material Deliveries

Equipment Use / Hours

Equipment Rentals

Daily Work Force **No.**

Superintendent	
Bricklayers	
Carpenters	
Masons	
Electricians	
Iron Workers	
Plumbers	
Others	
Total	

DAILY WORK LOG

7 AM

8 AM

9 AM

10 AM

11 AM

12 NOON

1 PM

2 PM

3 PM

4 PM

5 PM

6 PM

Delays / Problems

Schedule Updates / Progress

Extra Work / Authorized by

Supervisor's Signature ______________________

JOB NAME: ______________________ DATE: ______________

CONTRACTOR: ______________________ Weather: AM ________ PM ________

DAILY WORK LOG

Expenses / Materials

Material Deliveries

Equipment Use / Hours

Equipment Rentals

Daily Work Force No.

Superintendent	______
Bricklayers	______
Carpenters	______
Masons	______
Electricians	______
Iron Workers	______
Plumbers	______
Others	______
__________	______
Total	______

7 AM

8 AM

9 AM

10 AM

11 AM

12 NOON

1 PM

2 PM

3 PM

4 PM

5 PM

6 PM

Delays / Problems

Schedule Updates / Progress

Extra Work / Authorized by

Supervisor's Signature ______________________

JOB NAME: ______________________ DATE: ______________

CONTRACTOR: ______________________ Weather: AM ______ PM ______

Expenses / Materials

Material Deliveries

Equipment Use / Hours

Equipment Rentals

Daily Work Force No.

Superintendent	______
Bricklayers	______
Carpenters	______
Masons	______
Electricians	______
Iron Workers	______
Plumbers	______
Others	______
______________	______
Total	______

DAILY WORK LOG

7 AM

8 AM

9 AM

10 AM

11 AM

12 NOON

1 PM

2 PM

3 PM

4 PM

5 PM

6 PM

Delays / Problems

Schedule Updates / Progress

Extra Work / Authorized by

Supervisor's Signature ______________________________

JOB NAME: ____________________ DATE: __________

CONTRACTOR: ____________________ Weather: AM ______ PM ______

DAILY WORK LOG

Expenses / Materials

Material Deliveries

Equipment Use / Hours

Equipment Rentals

Daily Work Force **No.**

Superintendent	____
Bricklayers	____
Carpenters	____
Masons	____
Electricians	____
Iron Workers	____
Plumbers	____
Others	____
________	____
Total	____

7 AM

8 AM

9 AM

10 AM

11 AM

12 NOON

1 PM

2 PM

3 PM

4 PM

5 PM

6 PM

Delays / Problems

Schedule Updates / Progress

Extra Work / Authorized by

Supervisor's Signature ____________________

JOB NAME: ______________________ DATE: ______________

CONTRACTOR: ______________________ Weather: AM ______ PM ______

DAILY WORK LOG

Expenses / Materials

____

____

____

____

____

Material Deliveries

____

____

____

____

____

Equipment Use / Hours

____

____

____

____

____

Equipment Rentals

____

____

____

____

____

Daily Work Force	**No.**
Superintendent	______
Bricklayers	______
Carpenters	______
Masons	______
Electricians	______
Iron Workers	______
Plumbers	______
Others	______
______________	______
Total	______

7 AM ______________________________

____

8 AM ______________________________

____

9 AM ______________________________

____

10 AM ______________________________

____

11 AM ______________________________

____

12 NOON ______________________________

____

1 PM ______________________________

____

2 PM ______________________________

____

3 PM ______________________________

____

4 PM ______________________________

____

5 PM ______________________________

____

6 PM ______________________________

____

Delays / Problems ______________________________

____

____

Schedule Updates / Progress ______________________________

____

____

Extra Work / Authorized by ______________________________

____

Supervisor's Signature ______________________________

JOB NAME: ______________________ DATE: ______________

CONTRACTOR: ______________________ Weather: AM ______ PM ______

Expenses / Materials

Material Deliveries

Equipment Use / Hours

Equipment Rentals

Daily Work Force **No.**

Superintendent ______

Bricklayers ______

Carpenters ______

Masons ______

Electricians ______

Iron Workers ______

Plumbers ______

Others ______

______ ______

Total ______

DAILY WORK LOG

7 AM

8 AM

9 AM

10 AM

11 AM

12 NOON

1 PM

2 PM

3 PM

4 PM

5 PM

6 PM

Delays / Problems

Schedule Updates / Progress

Extra Work / Authorized by

Supervisor's Signature ______________________

JOB NAME: ______________________ DATE: ______________

CONTRACTOR: ______________________ Weather: AM ______ PM ______

DAILY WORK LOG

Expenses / Materials

____

____

____

____

____

Material Deliveries

____

____

____

____

____

Equipment Use / Hours

____

____

____

____

____

Equipment Rentals

____

____

____

____

____

Daily Work Force	**No.**
Superintendent	______
Bricklayers	______
Carpenters	______
Masons	______
Electricians	______
Iron Workers	______
Plumbers	______
Others	______
____________	______
Total	______

7 AM ______________________________

____

8 AM ______________________________

____

9 AM ______________________________

____

10 AM ______________________________

____

11 AM ______________________________

____

12 NOON ______________________________

____

1 PM ______________________________

____

2 PM ______________________________

____

3 PM ______________________________

____

4 PM ______________________________

____

5 PM ______________________________

____

6 PM ______________________________

____

Delays / Problems ______________________________

____

____

Schedule Updates / Progress ______________________________

____

____

Extra Work / Authorized by ______________________________

____

Supervisor's Signature ______________________________

JOB NAME: ______________________ DATE: __________

CONTRACTOR: ______________________ Weather: AM ______ PM ______

DAILY WORK LOG

Expenses / Materials

Material Deliveries

Equipment Use / Hours

Equipment Rentals

Daily Work Force **No.**

Superintendent	____
Bricklayers	____
Carpenters	____
Masons	____
Electricians	____
Iron Workers	____
Plumbers	____
Others	____
________	____
Total	____

7 AM

8 AM

9 AM

10 AM

11 AM

12 NOON

1 PM

2 PM

3 PM

4 PM

5 PM

6 PM

Delays / Problems

Schedule Updates / Progress

Extra Work / Authorized by

Supervisor's Signature ______________________

JOB NAME: ______________________ DATE: ______________

CONTRACTOR: ______________________ Weather: AM ______ PM ______

DAILY WORK LOG

Expenses / Materials

Material Deliveries

Equipment Use / Hours

Equipment Rentals

Daily Work Force **No.**

Superintendent ____

Bricklayers ____

Carpenters ____

Masons ____

Electricians ____

Iron Workers ____

Plumbers ____

Others ____

______________ ____

Total ____

7 AM

8 AM

9 AM

10 AM

11 AM

12 NOON

1 PM

2 PM

3 PM

4 PM

5 PM

6 PM

Delays / Problems

Schedule Updates / Progress

Extra Work / Authorized by

Supervisor's Signature ______________________

JOB NAME: ____________________ DATE: ____________

CONTRACTOR: ____________________ Weather: AM ______ PM ______

DAILY WORK LOG

Expenses / Materials

Material Deliveries

Equipment Use / Hours

Equipment Rentals

Daily Work Force **No.**

Superintendent ______

Bricklayers ______

Carpenters ______

Masons ______

Electricians ______

Iron Workers ______

Plumbers ______

Others ______

____________ ______

Total ______

7 AM

8 AM

9 AM

10 AM

11 AM

12 NOON

1 PM

2 PM

3 PM

4 PM

5 PM

6 PM

Delays / Problems

Schedule Updates / Progress

Extra Work / Authorized by

Supervisor's Signature ____________________

JOB NAME: ______________________ DATE: ______________

CONTRACTOR: ______________________ Weather: AM ______ PM ______

DAILY WORK LOG

Expenses / Materials

____

____

____

____

____

Material Deliveries

____

____

____

____

____

Equipment Use / Hours

____

____

____

____

____

Equipment Rentals

____

____

____

____

____

Daily Work Force	**No.**
Superintendent	______
Bricklayers	______
Carpenters	______
Masons	______
Electricians	______
Iron Workers	______
Plumbers	______
Others	______
______________	______
Total	______

7 AM ______________________________

____

8 AM ______________________________

____

9 AM ______________________________

____

10 AM ______________________________

____

11 AM ______________________________

____

12 NOON ______________________________

____

1 PM ______________________________

____

2 PM ______________________________

____

3 PM ______________________________

____

4 PM ______________________________

____

5 PM ______________________________

____

6 PM ______________________________

____

Delays / Problems ______________________________

____

____

Schedule Updates / Progress ______________________________

____

____

Extra Work / Authorized by ______________________________

____

Supervisor's Signature ______________________________

JOB NAME: ______________________ DATE: ____________

CONTRACTOR: ______________________ Weather: AM ________ PM ________

DAILY WORK LOG

Expenses / Materials

Material Deliveries

Equipment Use / Hours

Equipment Rentals

Daily Work Force **No.**

Daily Work Force	No.
Superintendent	____
Bricklayers	____
Carpenters	____
Masons	____
Electricians	____
Iron Workers	____
Plumbers	____
Others	____
________	____
Total	____

7 AM

8 AM

9 AM

10 AM

11 AM

12 NOON

1 PM

2 PM

3 PM

4 PM

5 PM

6 PM

Delays / Problems

Schedule Updates / Progress

Extra Work / Authorized by

Supervisor's Signature ______________________

JOB NAME: ______________________ DATE: ____________

CONTRACTOR: ______________________ Weather: AM ______ PM ______

Expenses / Materials

Material Deliveries

Equipment Use / [illegible]s

Equipment R[illegible]

Dail[illegible]k Force **No.**

Su[illegible]ndent ______

B[illegible]ers ______

[illegible]nters ______

[illegible]sons ______

[illegible]ectricians ______

Iron Workers ______

Plumbers ______

Others ______

______ ______

Total ______

DAILY WORK LOG

7 AM

8 AM

9 AM

10 AM

11 AM

12 NOON

1 PM

2 PM

3 PM

4 PM

5 PM

6 PM

Delays / Problems

Schedule Updates / Progress

Extra Work / Authorized b[illegible]

Supervisor's Signature ______________________

JOB NAME: ______________________ DATE: ______________

CONTRACTOR: ______________________ Weather: AM ______ PM ______

DAILY WORK LOG

Expenses / Materials

____

Material Deliveries

____

Equipment Use / Hours

____

Equipment Rentals

____

Daily Work Force	**No.**
Superintendent	
Bricklayers	
Carpenters	
Masons	
Electricians	
Iron Workers	
Plumbers	
Others	
Total	

7 AM

8 AM

9 AM

10 AM

11 AM

12 NOON

1 PM

2 PM

3 PM

4 PM

5 PM

6 PM

Delays / Problems

Schedule Updates / Progress

Extra Work / Authorized by

Supervisor's Signature ______________________

JOB NAME: ______________________ DATE: ______________

CONTRACTOR: ______________________ Weather: AM ______ PM ______

DAILY WORK LOG

Expenses / Materials

Material Deliveries

Equipment Use / Hours

Equipment Rentals

Daily Work Force	**No.**
Superintendent	
Bricklayers	
Carpenters	
Masons	
Electricians	
Iron Workers	
Plumbers	
Others	
Total	

7 AM

8 AM

9 AM

10 AM

11 AM

12 NOON

1 PM

2 PM

3 PM

4 PM

5 PM

6 PM

Delays / Problems

Schedule Updates / Progress

Extra Work / Authorized by

Supervisor's Signature ______________________

JOB NAME: ____________________ DATE: ____________

CONTRACTOR: ____________________ Weather: AM ________ PM ________

DAILY WORK LOG

Expenses / Materials

Material Deliveries

Equipment Use / Hours

Equipment Rentals

Daily Work Force	No.
Superintendent	
Bricklayers	
Carpenters	
Masons	
Electricians	
Iron Workers	
Plumbers	
Others	
Total	

7 AM

8 AM

9 AM

10 AM

11 AM

12 NOON

1 PM

2 PM

3 PM

4 PM

5 PM

6 PM

Delays / Problems

Schedule Updates / Progress

Extra Work / Authorized by

Supervisor's Signature ____________________

JOB NAME: ______________________ DATE: ____________

CONTRACTOR: ______________________ Weather: AM ______ PM ______

DAILY WORK LOG

Expenses / Materials

____

Material Deliveries

____

Equipment Use / Hours

____

Equipment Rentals

____

Daily Work Force **No.**

Superintendent ______

Bricklayers ______

Carpenters ______

Masons ______

Electricians ______

Iron Workers ______

Plumbers ______

Others ______

__________ ______

Total ______

7 AM ______________________

8 AM ______________________

9 AM ______________________

10 AM ______________________

11 AM ______________________

12 NOON ______________________

1 PM ______________________

2 PM ______________________

3 PM ______________________

4 PM ______________________

5 PM ______________________

6 PM ______________________

Delays / Problems ______________________

Schedule Updates / Progress ______________________

Extra Work / Authorized by ______________________

Supervisor's Signature ______________________

JOB NAME: ______________________________ DATE: ______________

CONTRACTOR: ______________________________ Weather: AM ________ PM ________

Expenses / Materials	**DAILY WORK LOG**
	7 AM
	8 AM
Material Deliveries	9 AM
	10 AM
	11 AM
Equipment Use / Hours	12 NOON
	1 PM
	2 PM
Equipment Rentals	3 PM
	4 PM
	5 PM
Daily Work Force **No.**	6 PM
Superintendent ______	
Bricklayers ______	
Carpenters ______	Delays / Problems
Masons ______	
Electricians ______	
Iron Workers ______	Schedule Updates / Progress
Plumbers ______	
Others ______	
____________ ______	Extra Work / Authorized by
Total ______	

Supervisor's Signature ______________________________

JOB NAME: ______________________ DATE: ______________

CONTRACTOR: ______________________ Weather: AM ______ PM ______

Expenses / Materials

____

____

____

____

____

Material Deliveries

____

____

____

____

____

Equipment Use / Hours

____

____

____

____

____

Equipment Rentals

____

____

____

____

____

Daily Work Force	**No.**
Superintendent	______
Bricklayers	______
Carpenters	______
Masons	______
Electricians	______
Iron Workers	______
Plumbers	______
Others	______
______________	______
Total	______

DAILY WORK LOG

7 AM ______________________________

____

8 AM ______________________________

____

9 AM ______________________________

____

10 AM ______________________________

____

11 AM ______________________________

____

12 NOON ______________________________

____

1 PM ______________________________

____

2 PM ______________________________

____

3 PM ______________________________

____

4 PM ______________________________

____

5 PM ______________________________

____

6 PM ______________________________

____

Delays / Problems ______________________________

____

____

Schedule Updates / Progress ______________________________

____

____

Extra Work / Authorized by ______________________________

____

Supervisor's Signature ______________________________

JOB NAME: ______________________ DATE: ______________

CONTRACTOR: ______________________ Weather: AM ______ PM ______

DAILY WORK LOG

Expenses / Materials

Material Deliveries

Equipment Use / Hours

Equipment Rentals

Daily Work Force **No.**

Superintendent ______

Bricklayers ______

Carpenters ______

Masons ______

Electricians ______

Iron Workers ______

Plumbers ______

Others ______

______________ ______

Total ______

7 AM

8 AM

9 AM

10 AM

11 AM

12 NOON

1 PM

2 PM

3 PM

4 PM

5 PM

6 PM

Delays / Problems

Schedule Updates / Progress

Extra Work / Authorized by

Supervisor's Signature ______________________

JOB NAME: ______________________ DATE: ______________

CONTRACTOR: ______________________ Weather: AM ______ PM ______

Expenses / Materials

____

Material Deliveries

____

Equipment Use / Hours

____

Equipment Rentals

____

Daily Work Force	**No.**
Superintendent	______
Bricklayers	______
Carpenters	______
Masons	______
Electricians	______
Iron Workers	______
Plumbers	______
Others	______
______________	______
Total	______

DAILY WORK LOG

7 AM ______________

8 AM ______________

9 AM ______________

10 AM ______________

11 AM ______________

12 NOON ______________

1 PM ______________

2 PM ______________

3 PM ______________

4 PM ______________

5 PM ______________

6 PM ______________

Delays / Problems ______________

Schedule Updates / Progress ______________

Extra Work / Authorized by ______________

Supervisor's Signature ______________________

JOB NAME: ______________________ DATE: ____________

CONTRACTOR: ______________________ Weather: AM ________ PM ________

DAILY WORK LOG

Expenses / Materials

Material Deliveries

Equipment Use / Hours

Equipment Rentals

Daily Work Force **No.**

Superintendent	____
Bricklayers	____
Carpenters	____
Masons	____
Electricians	____
Iron Workers	____
Plumbers	____
Others	____
________	____
Total	____

7 AM

8 AM

9 AM

10 AM

11 AM

12 NOON

1 PM

2 PM

3 PM

4 PM

5 PM

6 PM

Delays / Problems

Schedule Updates / Progress

Extra Work / Authorized by

Supervisor's Signature ______________________

JOB NAME: ______________________ DATE: ______________

CONTRACTOR: ______________________ Weather: AM ______ PM ______

DAILY WORK LOG

Expenses / Materials	7 AM
	8 AM
Material Deliveries	9 AM
	10 AM
	11 AM
Equipment Use / Hours	12 NOON
	1 PM
	2 PM
Equipment Rentals	3 PM
	4 PM
	5 PM
Daily Work Force **No.**	6 PM
Superintendent ______	
Bricklayers ______	
Carpenters ______	Delays / Problems
Masons ______	
Electricians ______	
Iron Workers ______	Schedule Updates / Progress
Plumbers ______	
Others ______	
______ ______	Extra Work / Authorized by
Total ______	

Supervisor's Signature ______________________

JOB NAME: ______________________ DATE: ______________

CONTRACTOR: ______________________ Weather: AM ________ PM ________

Expenses / Materials

Material Deliveries

Equipment Use / Hours

Equipment Rentals

Daily Work Force No.

Superintendent ______

Bricklayers ______

Carpenters ______

Masons ______

Electricians ______

Iron Workers ______

Plumbers ______

Others ______

______________ ______

Total ______

DAILY WORK LOG

7 AM

8 AM

9 AM

10 AM

11 AM

12 NOON

1 PM

2 PM

3 PM

4 PM

5 PM

6 PM

Delays / Problems

Schedule Updates / Progress

Extra Work / Authorized by

Supervisor's Signature ______________________________

JOB NAME: ______________________ DATE: ______________

CONTRACTOR: ______________________ Weather: AM ______ PM ______

Expenses / Materials

____

____

____

____

____

Material Deliveries

____

____

____

____

____

Equipment Use / Hours

____

____

____

____

____

Equipment Rentals

____

____

____

____

____

Daily Work Force	**No.**
Superintendent	______
Bricklayers	______
Carpenters	______
Masons	______
Electricians	______
Iron Workers	______
Plumbers	______
Others	______
______________	______
Total	______

DAILY WORK LOG

7 AM ______________________________________

____

8 AM ______________________________________

____

9 AM ______________________________________

____

10 AM ______________________________________

____

11 AM ______________________________________

____

12 NOON ______________________________________

____

1 PM ______________________________________

____

2 PM ______________________________________

____

3 PM ______________________________________

____

4 PM ______________________________________

____

5 PM ______________________________________

____

6 PM ______________________________________

____

Delays / Problems ______________________________________

____

____

Schedule Updates / Progress ______________________________________

____

____

Extra Work / Authorized by ______________________________________

____

Supervisor's Signature ______________________________________

JOB NAME: ____________ DATE: ____________

CONTRACTOR: ____________ Weather: AM ______ PM ______

DAILY WORK LOG

Expenses / Materials

Material Deliveries

Equipment Use / Hours

Equipment Rentals

Daily Work Force **No.**

Superintendent ______

Bricklayers ______

Carpenters ______

Masons ______

Electricians ______

Iron Workers ______

Plumbers ______

Others ______

______ ______

Total ______

7 AM

8 AM

9 AM

10 AM

11 AM

12 NOON

1 PM

2 PM

3 PM

4 PM

5 PM

6 PM

Delays / Problems

Schedule Updates / Progress

Extra Work / Authorized by

Supervisor's Signature ____________

JOB NAME: ____________________ DATE: ____________

CONTRACTOR: ____________________ Weather: AM ______ PM ______

DAILY WORK LOG

Expenses / Materials

Material Deliveries

Equipment Use / Hours

Equipment Rentals

Daily Work Force **No.**

Superintendent	____
Bricklayers	____
Carpenters	____
Masons	____
Electricians	____
Iron Workers	____
Plumbers	____
Others	____
________	____
Total	____

7 AM

8 AM

9 AM

10 AM

11 AM

12 NOON

1 PM

2 PM

3 PM

4 PM

5 PM

6 PM

Delays / Problems

Schedule Updates / Progress

Extra Work / Authorized by

Supervisor's Signature ______________________________

JOB NAME: ______________________ DATE: ____________

CONTRACTOR: ______________________ Weather: AM ______ PM ______

DAILY WORK LOG

Expenses / Materials

Material Deliveries

Equipment Use / Hours

Equipment Rentals

Daily Work Force	No.
Superintendent	
Bricklayers	
Carpenters	
Masons	
Electricians	
Iron Workers	
Plumbers	
Others	
Total	

7 AM

8 AM

9 AM

10 AM

11 AM

12 NOON

1 PM

2 PM

3 PM

4 PM

5 PM

6 PM

Delays / Problems

Schedule Updates / Progress

Extra Work / Authorized by

Supervisor's Signature ______________________

JOB NAME: ______________________ DATE: ______________

CONTRACTOR: ______________________ Weather: AM ______ PM ______

DAILY WORK LOG

Expenses / Materials

____

Material Deliveries

____

Equipment Use / Hours

____

Equipment Rentals

____

Daily Work Force **No.**

Superintendent	______
Bricklayers	______
Carpenters	______
Masons	______
Electricians	______
Iron Workers	______
Plumbers	______
Others	______
______________	______
Total	______

7 AM ______________________

8 AM ______________________

9 AM ______________________

10 AM ______________________

11 AM ______________________

12 NOON ______________________

1 PM ______________________

2 PM ______________________

3 PM ______________________

4 PM ______________________

5 PM ______________________

6 PM ______________________

Delays / Problems ______________________

Schedule Updates / Progress ______________________

Extra Work / Authorized by ______________________

Supervisor's Signature ______________________

JOB NAME: ______________________ DATE: ______________

CONTRACTOR: ______________________ Weather: AM ______ PM ______

DAILY WORK LOG

Expenses / Materials

Material Deliveries

Equipment Use / Hours

Equipment Rentals

Daily Work Force **No.**

Superintendent	____
Bricklayers	____
Carpenters	____
Masons	____
Electricians	____
Iron Workers	____
Plumbers	____
Others	____
________	____
Total	____

7 AM

8 AM

9 AM

10 AM

11 AM

12 NOON

1 PM

2 PM

3 PM

4 PM

5 PM

6 PM

Delays / Problems

Schedule Updates / Progress

Extra Work / Authorized by

Supervisor's Signature ______________________

JOB NAME: ______________________ DATE: ______________

CONTRACTOR: ______________________ Weather: AM ______ PM ______

Expenses / Materials

Material Deliveries

Equipment Use / Hours

Equipment Rentals

Daily Work Force **No.**

Superintendent	
Bricklayers	
Carpenters	
Masons	
Electricians	
Iron Workers	
Plumbers	
Others	
Total	

DAILY WORK LOG

7 AM

8 AM

9 AM

10 AM

11 AM

12 NOON

1 PM

2 PM

3 PM

4 PM

5 PM

6 PM

Delays / Problems

Schedule Updates / Progress

Extra Work / Authorized by

Supervisor's Signature ______________________

JOB NAME: ______________________ DATE: ______________

CONTRACTOR: ______________________ Weather: AM ________ PM ________

DAILY WORK LOG

Expenses / Materials

____

____

____

____

____

Material Deliveries

____

____

____

____

____

Equipment Use / Hours

____

____

____

____

____

Equipment Rentals

____

____

____

____

____

Daily Work Force	**No.**
Superintendent	____
Bricklayers	____
Carpenters	____
Masons	____
Electricians	____
Iron Workers	____
Plumbers	____
Others	____
____________	____
Total	____

7 AM ______________________________

____

8 AM ______________________________

____

9 AM ______________________________

____

10 AM ______________________________

____

11 AM ______________________________

____

12 NOON ______________________________

____

1 PM ______________________________

____

2 PM ______________________________

____

3 PM ______________________________

____

4 PM ______________________________

____

5 PM ______________________________

____

6 PM ______________________________

____

Delays / Problems ______________________________

____

____

Schedule Updates / Progress ______________________________

____

____

Extra Work / Authorized by ______________________________

____

Supervisor's Signature ______________________________

JOB NAME: ______________________ DATE: ______________

CONTRACTOR: ______________________ Weather: AM ______ PM ______

DAILY WORK LOG

Expenses / Materials

____

____

____

____

____

Material Deliveries

____

____

____

____

____

Equipment Use / Hours

____

____

____

____

____

Equipment Rentals

____

____

____

____

____

Daily Work Force **No.**

Superintendent	______
Bricklayers	______
Carpenters	______
Masons	______
Electricians	______
Iron Workers	______
Plumbers	______
Others	______
____________	______
Total	______

7 AM ______________________________

____

8 AM ______________________________

____

9 AM ______________________________

____

10 AM ______________________________

____

11 AM ______________________________

____

12 NOON ______________________________

____

1 PM ______________________________

____

2 PM ______________________________

____

3 PM ______________________________

____

4 PM ______________________________

____

5 PM ______________________________

____

6 PM ______________________________

____

Delays / Problems ______________________________

____

____

Schedule Updates / Progress ______________________________

____

____

Extra Work / Authorized by ______________________________

____

Supervisor's Signature ______________________________

JOB NAME: ______________________ DATE: ____________

CONTRACTOR: ______________________ Weather: AM ________ PM ________

Expenses / Materials

____

____

____

____

____

Material Deliveries

____

____

____

____

____

Equipment Use / Hours

____

____

____

____

____

Equipment Rentals

____

____

____

____

____

Daily Work Force	**No.**
Superintendent	____
Bricklayers	____
Carpenters	____
Masons	____
Electricians	____
Iron Workers	____
Plumbers	____
Others	____
____________	____
Total	____

DAILY WORK LOG

7 AM ______________________

8 AM ______________________

9 AM ______________________

10 AM ______________________

11 AM ______________________

12 NOON ______________________

1 PM ______________________

2 PM ______________________

3 PM ______________________

4 PM ______________________

5 PM ______________________

6 PM ______________________

Delays / Problems ______________________

Schedule Updates / Progress ______________________

Extra Work / Authorized by ______________________

Supervisor's Signature ______________________

JOB NAME: ______________________ DATE: ______________

CONTRACTOR: ______________________ Weather: AM ______ PM ______

Expenses / Materials

Material Deliveries

Equipment Use / Hours

Equipment Rentals

Daily Work Force **No.**

Superintendent ______

Bricklayers ______

Carpenters ______

Masons ______

Electricians ______

Iron Workers ______

Plumbers ______

Others ______

______________ ______

Total ______

DAILY WORK LOG

7 AM

8 AM

9 AM

10 AM

11 AM

12 NOON

1 PM

2 PM

3 PM

4 PM

5 PM

6 PM

Delays / Problems

Schedule Updates / Progress

Extra Work / Authorized by

Supervisor's Signature ______________________________

JOB NAME: ______________________ DATE: ____________

CONTRACTOR: ______________________ Weather: AM ______ PM ______

DAILY WORK LOG

Expenses / Materials

Material Deliveries

Equipment Use / Hours

Equipment Rentals

Daily Work Force **No.**

Superintendent	____
Bricklayers	____
Carpenters	____
Masons	____
Electricians	____
Iron Workers	____
Plumbers	____
Others	____
________	____
Total	____

7 AM

8 AM

9 AM

10 AM

11 AM

12 NOON

1 PM

2 PM

3 PM

4 PM

5 PM

6 PM

Delays / Problems

Schedule Updates / Progress

Extra Work / Authorized by

Supervisor's Signature ______________________

JOB NAME: ______________________ DATE: ______________

CONTRACTOR: ______________________ Weather: AM ______ PM ______

DAILY WORK LOG

Expenses / Materials

Material Deliveries

Equipment Use / Hours

Equipment Rentals

Daily Work Force	**No.**
Superintendent	
Bricklayers	
Carpenters	
Masons	
Electricians	
Iron Workers	
Plumbers	
Others	
Total	

7 AM

8 AM

9 AM

10 AM

11 AM

12 NOON

1 PM

2 PM

3 PM

4 PM

5 PM

6 PM

Delays / Problems

Schedule Updates / Progress

Extra Work / Authorized by

Supervisor's Signature ______________________

JOB NAME: ______________________ DATE: ______________

CONTRACTOR: ______________________ Weather: AM ______ PM ______

Expenses / Materials

DAILY WORK LOG

7 AM

8 AM

Material Deliveries

9 AM

10 AM

11 AM

Equipment Use / Hours

12 NOON

1 PM

2 PM

Equipment Rentals

3 PM

4 PM

5 PM

Daily Work Force No.

Superintendent	______
Bricklayers	______
Carpenters	______
Masons	______
Electricians	______
Iron Workers	______
Plumbers	______
Others	______
______________	______
Total	______

6 PM

Delays / Problems

Schedule Updates / Progress

Extra Work / Authorized by

Supervisor's Signature ______________________________

JOB NAME: ______________________ DATE: ______________

CONTRACTOR: ______________________ Weather: AM ________ PM ________

DAILY WORK LOG

Expenses / Materials	
	7 AM
	8 AM
Material Deliveries	9 AM
	10 AM
	11 AM
Equipment Use / Hours	
	12 NOON
	1 PM
Equipment Rentals	2 PM
	3 PM
	4 PM
Daily Work Force **No.**	5 PM
Superintendent ______	6 PM
Bricklayers ______	
Carpenters ______	Delays / Problems
Masons ______	
Electricians ______	
Iron Workers ______	Schedule Updates / Progress
Plumbers ______	
Others ______	
____________ ______	Extra Work / Authorized by
Total ______	

Supervisor's Signature ______________________________

JOB NAME: ______________________ DATE: ____________

CONTRACTOR: ______________________ Weather: AM ______ PM ______

DAILY WORK LOG

Expenses / Materials

Material Deliveries

Equipment Use / Hours

Equipment Rentals

Daily Work Force **No.**

Superintendent ______

Bricklayers ______

Carpenters ______

Masons ______

Electricians ______

Iron Workers ______

Plumbers ______

Others ______

______________ ______

Total ______

7 AM

8 AM

9 AM

10 AM

11 AM

12 NOON

1 PM

2 PM

3 PM

4 PM

5 PM

6 PM

Delays / Problems

Schedule Updates / Progress

Extra Work / Authorized by

Supervisor's Signature ______________________

JOB NAME: ______________________ DATE: ______________

CONTRACTOR: ______________________ Weather: AM ______ PM ______

Expenses / Materials

Material Deliveries

Equipment Use / Hours

Equipment Rentals

Daily Work Force **No.**

Superintendent	______
Bricklayers	______
Carpenters	______
Masons	______
Electricians	______
Iron Workers	______
Plumbers	______
Others	______
______________	______
Total	______

DAILY WORK LOG

7 AM

8 AM

9 AM

10 AM

11 AM

12 NOON

1 PM

2 PM

3 PM

4 PM

5 PM

6 PM

Delays / Problems

Schedule Updates / Progress

Extra Work / Authorized by

Supervisor's Signature ______________________

JOB NAME: ____________________ DATE: ____________

CONTRACTOR: ____________________ Weather: AM ______ PM ______

DAILY WORK LOG

Expenses / Materials

Material Deliveries

Equipment Use / Hours

Equipment Rentals

Daily Work Force	**No.**
Superintendent	
Bricklayers	
Carpenters	
Masons	
Electricians	
Iron Workers	
Plumbers	
Others	
Total	

7 AM

8 AM

9 AM

10 AM

11 AM

12 NOON

1 PM

2 PM

3 PM

4 PM

5 PM

6 PM

Delays / Problems

Schedule Updates / Progress

Extra Work / Authorized by

Supervisor's Signature ____________________

JOB NAME: ______________________ DATE: ______________

CONTRACTOR: ______________________ Weather: AM ______ PM ______

Expenses / Materials

Material Deliveries

Equipment Use / Hours

Equipment Rentals

Daily Work Force **No.**

Superintendent	______
Bricklayers	______
Carpenters	______
Masons	______
Electricians	______
Iron Workers	______
Plumbers	______
Others	______
______________	______
Total	______

DAILY WORK LOG

7 AM

8 AM

9 AM

10 AM

11 AM

12 NOON

1 PM

2 PM

3 PM

4 PM

5 PM

6 PM

Delays / Problems

Schedule Updates / Progress

Extra Work / Authorized by

Supervisor's Signature ______________________

JOB NAME: ______________________ DATE: ______________

CONTRACTOR: ______________________ Weather: AM ________ PM ________

Expenses / Materials

DAILY WORK LOG

7 AM

8 AM

9 AM

Material Deliveries

10 AM

11 AM

Equipment Use / Hours

12 NOON

1 PM

2 PM

Equipment Rentals

3 PM

4 PM

5 PM

Daily Work Force **No.**

Superintendent ______

Bricklayers ______

Carpenters ______

Masons ______

Electricians ______

Iron Workers ______

Plumbers ______

Others ______

______________ ______

Total ______

6 PM

Delays / Problems

Schedule Updates / Progress

Extra Work / Authorized by

Supervisor's Signature ______________________________

JOB NAME: ______________________ DATE: ______________

CONTRACTOR: ______________________ Weather: AM ______ PM ______

DAILY WORK LOG

Expenses / Materials

____

____

____

____

____

Material Deliveries

____

____

____

____

____

Equipment Use / Hours

____

____

____

____

____

Equipment Rentals

____

____

____

____

____

Daily Work Force	**No.**
Superintendent	______
Bricklayers	______
Carpenters	______
Masons	______
Electricians	______
Iron Workers	______
Plumbers	______
Others	______
______________	______
Total	______

7 AM ____

____

8 AM ____

____

9 AM ____

____

10 AM ____

____

11 AM ____

____

12 NOON ____

____

1 PM ____

____

2 PM ____

____

3 PM ____

____

4 PM ____

____

5 PM ____

____

6 PM ____

____

Delays / Problems ____

____

____

Schedule Updates / Progress ____

____

____

Extra Work / Authorized by ____

____

Supervisor's Signature ____

JOB NAME: ______________________ DATE: ____________

CONTRACTOR: ______________________ Weather: AM ________ PM ________

DAILY WORK LOG

Expenses / Materials

Material Deliveries

Equipment Use / Hours

Equipment Rentals

Daily Work Force **No.**

Superintendent ______

Bricklayers ______

Carpenters ______

Masons ______

Electricians ______

Iron Workers ______

Plumbers ______

Others ______

______________ ______

Total ______

7 AM

8 AM

9 AM

10 AM

11 AM

12 NOON

1 PM

2 PM

3 PM

4 PM

5 PM

6 PM

Delays / Problems

Schedule Updates / Progress

Extra Work / Authorized by

Supervisor's Signature ______________________

JOB NAME: ______________________ DATE: ______________

CONTRACTOR: ______________________ Weather: AM ______ PM ______

Expenses / Materials

____

____

____

____

____

Material Deliveries

____

____

____

____

____

Equipment Use / Hours

____

____

____

____

____

Equipment Rentals

____

____

____

____

____

Daily Work Force	**No.**
Superintendent	______
Bricklayers	______
Carpenters	______
Masons	______
Electricians	______
Iron Workers	______
Plumbers	______
Others	______
______________	______
Total	______

DAILY WORK LOG

7 AM ______________________

8 AM ______________________

9 AM ______________________

10 AM ______________________

11 AM ______________________

12 NOON ______________________

1 PM ______________________

2 PM ______________________

3 PM ______________________

4 PM ______________________

5 PM ______________________

6 PM ______________________

Delays / Problems ______________________

Schedule Updates / Progress ______________________

Extra Work / Authorized by ______________________

Supervisor's Signature ______________________

JOB NAME: ______________________ DATE: ______________

CONTRACTOR: ______________________ Weather: AM ________ PM ________

Expenses / Materials

DAILY WORK LOG

7 AM

8 AM

9 AM

10 AM

11 AM

12 NOON

1 PM

2 PM

3 PM

4 PM

5 PM

6 PM

Delays / Problems

Schedule Updates / Progress

Extra Work / Authorized by

Material Deliveries

Equipment Use / Hours

Equipment Rentals

Daily Work Force	**No.**
Superintendent	______
Bricklayers	______
Carpenters	______
Masons	______
Electricians	______
Iron Workers	______
Plumbers	______
Others	______
____________	______
Total	______

Supervisor's Signature ______________________________

JOB NAME: ______________________ DATE: ______________

CONTRACTOR: ______________________ Weather: AM ______ PM ______

DAILY WORK LOG

Expenses / Materials

Material Deliveries

Equipment Use / Hours

Equipment Rentals

Daily Work Force	**No.**
Superintendent	
Bricklayers	
Carpenters	
Masons	
Electricians	
Iron Workers	
Plumbers	
Others	
Total	

7 AM

8 AM

9 AM

10 AM

11 AM

12 NOON

1 PM

2 PM

3 PM

4 PM

5 PM

6 PM

Delays / Problems

Schedule Updates / Progress

Extra Work / Authorized by

Supervisor's Signature ______________________

JOB NAME: ______________________ DATE: ______________

CONTRACTOR: ______________________ Weather: AM ______ PM ______

Expenses / Materials

Material Deliveries

Equipment Use / Hours

Equipment Rentals

Daily Work Force **No.**

Superintendent ______

Bricklayers ______

Carpenters ______

Masons ______

Electricians ______

Iron Workers ______

Plumbers ______

Others ______

______ ______

Total ______

DAILY WORK LOG

7 AM

8 AM

9 AM

10 AM

11 AM

12 NOON

1 PM

2 PM

3 PM

4 PM

5 PM

6 PM

Delays / Problems

Schedule Updates / Progress

Extra Work / Authorized by

Supervisor's Signature ______________________

JOB NAME: ______________________ DATE: ______________

CONTRACTOR: ______________________ Weather: AM ______ PM ______

DAILY WORK LOG

Expenses / Materials

Material Deliveries

Equipment Use / Hours

Equipment Rentals

Daily Work Force	**No.**
Superintendent	
Bricklayers	
Carpenters	
Masons	
Electricians	
Iron Workers	
Plumbers	
Others	
Total	

7 AM

8 AM

9 AM

10 AM

11 AM

12 NOON

1 PM

2 PM

3 PM

4 PM

5 PM

6 PM

Delays / Problems

Schedule Updates / Progress

Extra Work / Authorized by

Supervisor's Signature ______________________

JOB NAME: ______________________ DATE: ______________

CONTRACTOR: ______________________ Weather: AM ______ PM ______

Expenses / Materials

Material Deliveries

Equipment Use / Hours

Equipment Rentals

Daily Work Force	**No.**
Superintendent	
Bricklayers	
Carpenters	
Masons	
Electricians	
Iron Workers	
Plumbers	
Others	
Total	

DAILY WORK LOG

7 AM

8 AM

9 AM

10 AM

11 AM

12 NOON

1 PM

2 PM

3 PM

4 PM

5 PM

6 PM

Delays / Problems

Schedule Updates / Progress

Extra Work / Authorized by

Supervisor's Signature ______________________

JOB NAME: ______________________ DATE: ____________

CONTRACTOR: ______________________ Weather: AM ______ PM ______

DAILY WORK LOG

Expenses / Materials

Material Deliveries

Equipment Use / Hours

Equipment Rentals

Daily Work Force **No.**

Superintendent	______
Bricklayers	______
Carpenters	______
Masons	______
Electricians	______
Iron Workers	______
Plumbers	______
Others	______
____________	______
Total	______

7 AM

8 AM

9 AM

10 AM

11 AM

12 NOON

1 PM

2 PM

3 PM

4 PM

5 PM

6 PM

Delays / Problems

Schedule Updates / Progress

Extra Work / Authorized by

Supervisor's Signature ______________________

JOB NAME: ______________________ DATE: ____________

CONTRACTOR: ______________________ Weather: AM ______ PM ______

Expenses / Materials

Material Deliveries

Equipment Use / Hours

Equipment Rentals

Daily Work Force No.

Superintendent ____

Bricklayers ____

Carpenters ____

Masons ____

Electricians ____

Iron Workers ____

Plumbers ____

Others ____

__________ ____

Total ____

DAILY WORK LOG

7 AM

8 AM

9 AM

10 AM

11 AM

12 NOON

1 PM

2 PM

3 PM

4 PM

5 PM

6 PM

Delays / Problems

Schedule Updates / Progress

Extra Work / Authorized by

Supervisor's Signature ______________________

JOB NAME: ______________________ DATE: ______________

CONTRACTOR: ______________________ Weather: AM ______ PM ______

DAILY WORK LOG

Expenses / Materials

____

____

____

____

____

Material Deliveries

____

____

____

____

____

Equipment Use / Hours

____

____

____

____

____

Equipment Rentals

____

____

____

____

____

Daily Work Force	**No.**
Superintendent	______
Bricklayers	______
Carpenters	______
Masons	______
Electricians	______
Iron Workers	______
Plumbers	______
Others	______
______________	______
Total	______

7 AM ______________________________

____

8 AM ______________________________

____

9 AM ______________________________

____

10 AM ______________________________

____

11 AM ______________________________

____

12 NOON ______________________________

____

1 PM ______________________________

____

2 PM ______________________________

____

3 PM ______________________________

____

4 PM ______________________________

____

5 PM ______________________________

____

6 PM ______________________________

____

Delays / Problems ______________________________

____

____

Schedule Updates / Progress ______________________________

____

____

Extra Work / Authorized by ______________________________

____

Supervisor's Signature ______________________________

JOB NAME: ______________________ DATE: ____________

CONTRACTOR: ______________________ Weather: AM ______ PM ______

DAILY WORK LOG

Expenses / Materials

Material Deliveries

Equipment Use / Hours

Equipment Rentals

Daily Work Force **No.**

Superintendent	____
Bricklayers	____
Carpenters	____
Masons	____
Electricians	____
Iron Workers	____
Plumbers	____
Others	____
________	____
Total	____

7 AM

8 AM

9 AM

10 AM

11 AM

12 NOON

1 PM

2 PM

3 PM

4 PM

5 PM

6 PM

Delays / Problems

Schedule Updates / Progress

Extra Work / Authorized by

Supervisor's Signature ______________________

JOB NAME: ______________________ DATE: ______________

CONTRACTOR: ______________________ Weather: AM ______ PM ______

DAILY WORK LOG

Expenses / Materials	
______	7 AM ______
______	______
______	8 AM ______
______	______
Material Deliveries	9 AM ______
______	______
______	10 AM ______
______	______
______	11 AM ______
Equipment Use / Hours	______
______	12 NOON ______
______	______
______	1 PM ______
______	______
______	2 PM ______
Equipment Rentals	______
______	3 PM ______
______	______
______	4 PM ______
______	______
______	5 PM ______
	6 PM ______
	______
	Delays / Problems ______
	______
	______
	Schedule Updates / Progress ______
	______
	______
	Extra Work / Authorized by ______
	______

Daily Work Force	**No.**
Superintendent	______
Bricklayers	______
Carpenters	______
Masons	______
Electricians	______
Iron Workers	______
Plumbers	______
Others	______
______	______
Total	______

Supervisor's Signature ______________________________

JOB NAME: ______________________ DATE: ______________

CONTRACTOR: ______________________ Weather: AM ______ PM ______

DAILY WORK LOG

Expenses / Materials

Material Deliveries

Equipment Use / Hours

Equipment Rentals

Daily Work Force **No.**

Daily Work Force	**No.**
Superintendent	
Bricklayers	
Carpenters	
Masons	
Electricians	
Iron Workers	
Plumbers	
Others	
______	
Total	

7 AM

8 AM

9 AM

10 AM

11 AM

12 NOON

1 PM

2 PM

3 PM

4 PM

5 PM

6 PM

Delays / Problems

Schedule Updates / Progress

Extra Work / Authorized by

Supervisor's Signature ______________________

JOB NAME: ______________________ DATE: ______________

CONTRACTOR: ______________________ Weather: AM ______ PM ______

DAILY WORK LOG

Expenses / Materials

____

____

____

____

____

Material Deliveries

____

____

____

____

____

Equipment Use / Hours

____

____

____

____

____

Equipment Rentals

____

____

____

____

____

Daily Work Force	**No.**
Superintendent	______
Bricklayers	______
Carpenters	______
Masons	______
Electricians	______
Iron Workers	______
Plumbers	______
Others	______
______________	______
Total	______

7 AM ____

____

8 AM ____

____

9 AM ____

____

10 AM ____

____

11 AM ____

____

12 NOON ____

____

1 PM ____

____

2 PM ____

____

3 PM ____

____

4 PM ____

____

5 PM ____

____

6 PM ____

____

Delays / Problems ____

____

____

Schedule Updates / Progress ____

____

____

Extra Work / Authorized by ____

____

Supervisor's Signature ____

JOB NAME: ______________________ DATE: ____________

CONTRACTOR: ______________________ Weather: AM ______ PM ______

DAILY WORK LOG

Expenses / Materials

Material Deliveries

Equipment Use / Hours

Equipment Rentals

Daily Work Force **No.**

Superintendent	____
Bricklayers	____
Carpenters	____
Masons	____
Electricians	____
Iron Workers	____
Plumbers	____
Others	____
________	____
Total	____

7 AM

8 AM

9 AM

10 AM

11 AM

12 NOON

1 PM

2 PM

3 PM

4 PM

5 PM

6 PM

Delays / Problems

Schedule Updates / Progress

Extra Work / Authorized by

Supervisor's Signature ______________________

JOB NAME: ______________________ DATE: ______________

CONTRACTOR: ______________________ Weather: AM ______ PM ______

Expenses / Materials

____

____

____

____

____

Material Deliveries

____

____

____

____

____

Equipment Use / Hours

____

____

____

____

____

Equipment Rentals

____

____

____

____

____

Daily Work Force	**No.**
Superintendent	______
Bricklayers	______
Carpenters	______
Masons	______
Electricians	______
Iron Workers	______
Plumbers	______
Others	______
______________	______
Total	______

DAILY WORK LOG

7 AM ______________________________

____

8 AM ______________________________

____

9 AM ______________________________

____

10 AM ______________________________

____

11 AM ______________________________

____

12 NOON ______________________________

____

1 PM ______________________________

____

2 PM ______________________________

____

3 PM ______________________________

____

4 PM ______________________________

____

5 PM ______________________________

____

6 PM ______________________________

____

Delays / Problems ______________________________

____

____

Schedule Updates / Progress ______________________________

____

____

Extra Work / Authorized by ______________________________

____

Supervisor's Signature ______________________________

JOB NAME: ______________________ DATE: ____________

CONTRACTOR: ______________________ Weather: AM ______ PM ______

DAILY WORK LOG

Expenses / Materials

____

Material Deliveries

____

Equipment Use / Hours

____

Equipment Rentals

____

Daily Work Force	**No.**
Superintendent	______
Bricklayers	______
Carpenters	______
Masons	______
Electricians	______
Iron Workers	______
Plumbers	______
Others	______
____________	______
Total	______

7 AM ______________________

8 AM ______________________

9 AM ______________________

10 AM ______________________

11 AM ______________________

12 NOON ______________________

1 PM ______________________

2 PM ______________________

3 PM ______________________

4 PM ______________________

5 PM ______________________

6 PM ______________________

Delays / Problems ______________________

Schedule Updates / Progress ______________________

Extra Work / Authorized by ______________________

Supervisor's Signature ______________________

JOB NAME: ______________________ DATE: ______________

CONTRACTOR: ______________________ Weather: AM ______ PM ______

DAILY WORK LOG

Expenses / Materials

____

____

____

____

____

Material Deliveries

____

____

____

____

____

Equipment Use / Hours

____

____

____

____

____

Equipment Rentals

____

____

____

____

____

Daily Work Force	**No.**
Superintendent	______
Bricklayers	______
Carpenters	______
Masons	______
Electricians	______
Iron Workers	______
Plumbers	______
Others	______
______________	______
Total	______

7 AM ______________________

____

8 AM ______________________

____

9 AM ______________________

____

10 AM ______________________

____

11 AM ______________________

____

12 NOON ______________________

____

1 PM ______________________

____

2 PM ______________________

____

3 PM ______________________

____

4 PM ______________________

____

5 PM ______________________

____

6 PM ______________________

____

Delays / Problems ______________________

____

____

Schedule Updates / Progress ______________________

____

____

Extra Work / Authorized by ______________________

____

Supervisor's Signature ______________________

JOB NAME: ______________________ DATE: ______________

CONTRACTOR: ______________________ Weather: AM ________ PM ________

DAILY WORK LOG

Expenses / Materials

Material Deliveries

Equipment Use / Hours

Equipment Rentals

Daily Work Force **No.**

Daily Work Force	No.
Superintendent	
Bricklayers	
Carpenters	
Masons	
Electricians	
Iron Workers	
Plumbers	
Others	
Total	

7 AM

8 AM

9 AM

10 AM

11 AM

12 NOON

1 PM

2 PM

3 PM

4 PM

5 PM

6 PM

Delays / Problems

Schedule Updates / Progress

Extra Work / Authorized by

Supervisor's Signature ______________________

JOB NAME: ______________________ DATE: ______________

CONTRACTOR: ______________________ Weather: AM ______ PM ______

DAILY WORK LOG

Expenses / Materials	
	7 AM
	8 AM
Material Deliveries	9 AM
	10 AM
	11 AM
Equipment Use / Hours	
	12 NOON
	1 PM
	2 PM
Equipment Rentals	
	3 PM
	4 PM
	5 PM

Daily Work Force **No.**

Superintendent	______
Bricklayers	______
Carpenters	______
Masons	______
Electricians	______
Iron Workers	______
Plumbers	______
Others	______
______________	______
Total	______

6 PM

Delays / Problems

Schedule Updates / Progress

Extra Work / Authorized by

Supervisor's Signature ______________________________

JOB NAME: ____________________ DATE: ____________

CONTRACTOR: ____________________ Weather: AM ________ PM ________

Expenses / Materials

Material Deliveries

Equipment Use / Hours

Equipment Rentals

Daily Work Force **No.**

Superintendent ______

Bricklayers ______

Carpenters ______

Masons ______

Electricians ______

Iron Workers ______

Plumbers ______

Others ______

____________ ______

Total ______

DAILY WORK LOG

7 AM

8 AM

9 AM

10 AM

11 AM

12 NOON

1 PM

2 PM

3 PM

4 PM

5 PM

6 PM

Delays / Problems

Schedule Updates / Progress

Extra Work / Authorized by

Supervisor's Signature ____________________

JOB NAME: ______________________ DATE: ______________

CONTRACTOR: ______________________ Weather: AM ________ PM ________

DAILY WORK LOG

Expenses / Materials

____

____

____

____

____

Material Deliveries

____

____

____

____

____

Equipment Use / Hours

____

____

____

____

____

Equipment Rentals

____

____

____

____

____

Daily Work Force	**No.**
Superintendent	______
Bricklayers	______
Carpenters	______
Masons	______
Electricians	______
Iron Workers	______
Plumbers	______
Others	______
______________	______
Total	______

7 AM ______________________________

____

8 AM ______________________________

____

9 AM ______________________________

____

10 AM ______________________________

____

11 AM ______________________________

____

12 NOON ______________________________

____

1 PM ______________________________

____

2 PM ______________________________

____

3 PM ______________________________

____

4 PM ______________________________

____

5 PM ______________________________

____

6 PM ______________________________

____

Delays / Problems ______________________________

____

____

Schedule Updates / Progress ______________________________

____

____

Extra Work / Authorized by ______________________________

____

Supervisor's Signature ______________________________

JOB NAME: ______________________ DATE: ______________

CONTRACTOR: ______________________ Weather: AM ______ PM ______

DAILY WORK LOG

Expenses / Materials

____

____

____

____

____

Material Deliveries

____

____

____

____

____

Equipment Use / Hours

____

____

____

____

____

Equipment Rentals

____

____

____

____

____

Daily Work Force	**No.**
Superintendent	______
Bricklayers	______
Carpenters	______
Masons	______
Electricians	______
Iron Workers	______
Plumbers	______
Others	______
______________	______
Total	______

7 AM ______________________________

____

8 AM ______________________________

____

9 AM ______________________________

____

10 AM ______________________________

____

11 AM ______________________________

____

12 NOON ______________________________

____

1 PM ______________________________

____

2 PM ______________________________

____

3 PM ______________________________

____

4 PM ______________________________

____

5 PM ______________________________

____

6 PM ______________________________

____

Delays / Problems ______________________________

____

____

Schedule Updates / Progress ______________________________

____

____

Extra Work / Authorized by ______________________________

____

Supervisor's Signature ______________________________

JOB NAME: ____________________ DATE: ____________

CONTRACTOR: ____________________ Weather: AM ______ PM ______

Expenses / Materials

Material Deliveries

Equipment Use / Hours

Equipment Rentals

Daily Work Force	No.
Superintendent	
Bricklayers	
Carpenters	
Masons	
Electricians	
Iron Workers	
Plumbers	
Others	
Total	

DAILY WORK LOG

7 AM

8 AM

9 AM

10 AM

11 AM

12 NOON

1 PM

2 PM

3 PM

4 PM

5 PM

6 PM

Delays / Problems

Schedule Updates / Progress

Extra Work / Authorized by

Supervisor's Signature ____________________

JOB NAME: ______________________ DATE: ____________

CONTRACTOR: ______________________ Weather: AM ________ PM ________

DAILY WORK LOG

Expenses / Materials

Material Deliveries

Equipment Use / Hours

Equipment Rentals

Daily Work Force No.

Superintendent ____

Bricklayers ____

Carpenters ____

Masons ____

Electricians ____

Iron Workers ____

Plumbers ____

Others ____

________ ____

Total ____

7 AM

8 AM

9 AM

10 AM

11 AM

12 NOON

1 PM

2 PM

3 PM

4 PM

5 PM

6 PM

Delays / Problems

Schedule Updates / Progress

Extra Work / Authorized by

Supervisor's Signature ______________________

JOB NAME: ______________________ DATE: ______________

CONTRACTOR: ______________________ Weather: AM ______ PM ______

DAILY WORK LOG

Expenses / Materials

____

____

____

____

____

Material Deliveries

____

____

____

____

____

Equipment Use / Hours

____

____

____

____

____

Equipment Rentals

____

____

____

____

____

Daily Work Force	**No.**
Superintendent	______
Bricklayers	______
Carpenters	______
Masons	______
Electricians	______
Iron Workers	______
Plumbers	______
Others	______
____________	______
Total	______

7 AM ______________________

____

8 AM ______________________

____

9 AM ______________________

____

10 AM ______________________

____

11 AM ______________________

____

12 NOON ______________________

____

1 PM ______________________

____

2 PM ______________________

____

3 PM ______________________

____

4 PM ______________________

____

5 PM ______________________

____

6 PM ______________________

____

Delays / Problems ______________________

____

____

Schedule Updates / Progress ______________________

____

____

Extra Work / Authorized by ______________________

____

Supervisor's Signature ______________________

JOB NAME: ______________________ DATE: ____________

CONTRACTOR: ______________________ Weather: AM ________ PM ________

Expenses / Materials

Material Deliveries

Equipment Use / Hours

Equipment Rentals

Daily Work Force	**No.**
Superintendent	
Bricklayers	
Carpenters	
Masons	
Electricians	
Iron Workers	
Plumbers	
Others	
Total	

DAILY WORK LOG

7 AM

8 AM

9 AM

10 AM

11 AM

12 NOON

1 PM

2 PM

3 PM

4 PM

5 PM

6 PM

Delays / Problems

Schedule Updates / Progress

Extra Work / Authorized by

Supervisor's Signature ______________________

JOB NAME: ______________________ DATE: ______________

CONTRACTOR: ______________________ Weather: AM ______ PM ______

Expenses / Materials

____

Material Deliveries

____

Equipment Use / Hours

____

Equipment Rentals

____

Daily Work Force	**No.**
Superintendent	______
Bricklayers	______
Carpenters	______
Masons	______
Electricians	______
Iron Workers	______
Plumbers	______
Others	______
______________	______
Total	______

DAILY WORK LOG

7 AM ______________________

8 AM ______________________

9 AM ______________________

10 AM ______________________

11 AM ______________________

12 NOON ______________________

1 PM ______________________

2 PM ______________________

3 PM ______________________

4 PM ______________________

5 PM ______________________

6 PM ______________________

Delays / Problems ______________________

Schedule Updates / Progress ______________________

Extra Work / Authorized by ______________________

Supervisor's Signature ______________________

JOB NAME: ______________________ DATE: __________

CONTRACTOR: ______________________ Weather: AM ______ PM ______

DAILY WORK LOG

Expenses / Materials

Material Deliveries

Equipment Use / Hours

Equipment Rentals

Daily Work Force	**No.**
Superintendent	
Bricklayers	
Carpenters	
Masons	
Electricians	
Iron Workers	
Plumbers	
Others	
Total	

7 AM

8 AM

9 AM

10 AM

11 AM

12 NOON

1 PM

2 PM

3 PM

4 PM

5 PM

6 PM

Delays / Problems

Schedule Updates / Progress

Extra Work / Authorized by

Supervisor's Signature ______________________

JOB NAME: ______________________ DATE: ______________

CONTRACTOR: ______________________ Weather: AM ______ PM ______

DAILY WORK LOG

Expenses / Materials

____

____

____

____

____

Material Deliveries

____

____

____

____

____

Equipment Use / Hours

____

____

____

____

____

Equipment Rentals

____

____

____

____

____

Daily Work Force	**No.**
Superintendent	______
Bricklayers	______
Carpenters	______
Masons	______
Electricians	______
Iron Workers	______
Plumbers	______
Others	______
______________	______
Total	______

7 AM ______________________________

____

8 AM ______________________________

____

9 AM ______________________________

____

10 AM ______________________________

____

11 AM ______________________________

____

12 NOON ______________________________

____

1 PM ______________________________

____

2 PM ______________________________

____

3 PM ______________________________

____

4 PM ______________________________

____

5 PM ______________________________

____

6 PM ______________________________

____

Delays / Problems ______________________________

____

____

Schedule Updates / Progress ______________________________

____

____

Extra Work / Authorized by ______________________________

____

Supervisor's Signature ______________________________

JOB NAME: ______________________ DATE: __________

CONTRACTOR: ______________________ Weather: AM ______ PM ______

Expenses / Materials

DAILY WORK LOG

7 AM

8 AM

9 AM

10 AM

11 AM

12 NOON

1 PM

2 PM

3 PM

4 PM

5 PM

6 PM

Delays / Problems

Schedule Updates / Progress

Extra Work / Authorized by

Material Deliveries

Equipment Use / Hours

Equipment Rentals

Daily Work Force	**No.**
Superintendent	______
Bricklayers	______
Carpenters	______
Masons	______
Electricians	______
Iron Workers	______
Plumbers	______
Others	______
____________	______
Total	______

Supervisor's Signature ______________________

JOB NAME: ______________________ DATE: ______________

CONTRACTOR: ______________________ Weather: AM ______ PM ______

Expenses / Materials

____

____

____

____

____

Material Deliveries

____

____

____

____

____

Equipment Use / Hours

____

____

____

____

____

Equipment Rentals

____

____

____

____

____

Daily Work Force	**No.**
Superintendent	______
Bricklayers	______
Carpenters	______
Masons	______
Electricians	______
Iron Workers	______
Plumbers	______
Others	______
____________	______
Total	______

DAILY WORK LOG

7 AM ______________________________________

____

8 AM ______________________________________

____

9 AM ______________________________________

____

10 AM ______________________________________

____

11 AM ______________________________________

____

12 NOON ______________________________________

____

1 PM ______________________________________

____

2 PM ______________________________________

____

3 PM ______________________________________

____

4 PM ______________________________________

____

5 PM ______________________________________

____

6 PM ______________________________________

____

Delays / Problems ______________________________________

____

____

Schedule Updates / Progress ______________________________________

____

____

Extra Work / Authorized by ______________________________________

____

Supervisor's Signature ______________________________________

JOB NAME: ______________________________ DATE: ______________

CONTRACTOR: ______________________________ Weather: AM ________ PM ________

Expenses / Materials

____

____

____

____

____

Material Deliveries

____

____

____

____

____

Equipment Use / Hours

____

____

____

____

____

Equipment Rentals

____

____

____

____

____

Daily Work Force	**No.**
Superintendent	______
Bricklayers	______
Carpenters	______
Masons	______
Electricians	______
Iron Workers	______
Plumbers	______
Others	______
______________	______
Total	______

DAILY WORK LOG

7 AM ______________________________________

____

8 AM ______________________________________

____

9 AM ______________________________________

____

10 AM ______________________________________

____

11 AM ______________________________________

____

12 NOON ______________________________________

____

1 PM ______________________________________

____

2 PM ______________________________________

____

3 PM ______________________________________

____

4 PM ______________________________________

____

5 PM ______________________________________

____

6 PM ______________________________________

____

Delays / Problems ______________________________________

____

____

Schedule Updates / Progress ______________________________________

____

____

Extra Work / Authorized by ______________________________________

____

Supervisor's Signature ______________________________________

JOB NAME: ______________________ DATE: ____________

CONTRACTOR: ______________________ Weather: AM ______ PM ______

DAILY WORK LOG

Expenses / Materials

____

Material Deliveries

____

Equipment Use / Hours

____

Equipment Rentals

____

Daily Work Force **No.**

Superintendent	______
Bricklayers	______
Carpenters	______
Masons	______
Electricians	______
Iron Workers	______
Plumbers	______
Others	______
______________	______
Total	______

7 AM ______________________

8 AM ______________________

9 AM ______________________

10 AM ______________________

11 AM ______________________

12 NOON ______________________

1 PM ______________________

2 PM ______________________

3 PM ______________________

4 PM ______________________

5 PM ______________________

6 PM ______________________

Delays / Problems ______________________

Schedule Updates / Progress ______________________

Extra Work / Authorized by ______________________

Supervisor's Signature ______________________

JOB NAME: ____________________ DATE: ____________

CONTRACTOR: ____________________ Weather: AM ______ PM ______

Expenses / Materials

DAILY WORK LOG

7 AM

8 AM

Material Deliveries

9 AM

10 AM

11 AM

Equipment Use / Hours

12 NOON

1 PM

2 PM

Equipment Rentals

3 PM

4 PM

5 PM

Daily Work Force	**No.**
Superintendent	______
Bricklayers	______
Carpenters	______
Masons	______
Electricians	______
Iron Workers	______
Plumbers	______
Others	______
____________	______
Total	______

6 PM

Delays / Problems

Schedule Updates / Progress

Extra Work / Authorized by

Supervisor's Signature ______________________________

JOB NAME: ______________________ DATE: ______________

CONTRACTOR: ______________________ Weather: AM ______ PM ______

Expenses / Materials	**DAILY WORK LOG**
	7 AM
	8 AM
Material Deliveries	9 AM
	10 AM
	11 AM
Equipment Use / Hours	12 NOON
	1 PM
	2 PM
Equipment Rentals	3 PM
	4 PM
	5 PM
Daily Work Force **No.**	6 PM
Superintendent ______	
Bricklayers ______	
Carpenters ______	Delays / Problems
Masons ______	
Electricians ______	
Iron Workers ______	Schedule Updates / Progress
Plumbers ______	
Others ______	
____________ ______	Extra Work / Authorized by
Total ______	

Supervisor's Signature ______________________________

JOB NAME: ______________________ DATE: ______________

CONTRACTOR: ______________________ Weather: AM ______ PM ______

DAILY WORK LOG

Expenses / Materials

Material Deliveries

Equipment Use / Hours

Equipment Rentals

Daily Work Force	**No.**
Superintendent	
Bricklayers	
Carpenters	
Masons	
Electricians	
Iron Workers	
Plumbers	
Others	
Total	

7 AM

8 AM

9 AM

10 AM

11 AM

12 NOON

1 PM

2 PM

3 PM

4 PM

5 PM

6 PM

Delays / Problems

Schedule Updates / Progress

Extra Work / Authorized by

Supervisor's Signature ______________________

JOB NAME: ______________________ DATE: ______________

CONTRACTOR: ______________________ Weather: AM ______ PM ______

Expenses / Materials

____
____
____
____
____

Material Deliveries

____
____
____
____
____

Equipment Use / Hours

____
____
____
____
____

Equipment Rentals

____
____
____
____
____

Daily Work Force	**No.**
Superintendent	______
Bricklayers	______
Carpenters	______
Masons	______
Electricians	______
Iron Workers	______
Plumbers	______
Others	______
______________	______
Total	______

DAILY WORK LOG

7 AM ______________________

8 AM ______________________

9 AM ______________________

10 AM ______________________

11 AM ______________________

12 NOON ______________________

1 PM ______________________

2 PM ______________________

3 PM ______________________

4 PM ______________________

5 PM ______________________

6 PM ______________________

Delays / Problems ______________________

Schedule Updates / Progress ______________________

Extra Work / Authorized by ______________________

Supervisor's Signature ______________________

JOB NAME: ______________________ DATE: ____________

CONTRACTOR: ______________________ Weather: AM ______ PM ______

Expenses / Materials

Material Deliveries

Equipment Use / Hours

Equipment Rentals

Daily Work Force No.

Superintendent ______

Bricklayers ______

Carpenters ______

Masons ______

Electricians ______

Iron Workers ______

Plumbers ______

Others ______

__________ ______

Total ______

DAILY WORK LOG

7 AM

8 AM

9 AM

10 AM

11 AM

12 NOON

1 PM

2 PM

3 PM

4 PM

5 PM

6 PM

Delays / Problems

Schedule Updates / Progress

Extra Work / Authorized by

Supervisor's Signature ______________________

JOB NAME: ______________________ DATE: ____________

CONTRACTOR: ______________________ Weather: AM ______ PM ______

Expenses / Materials

____
____
____
____
____

Material Deliveries

____
____
____
____
____

Equipment Use / Hours

____
____
____
____
____

Equipment Rentals

____
____
____
____
____

Daily Work Force **No.**

Superintendent ______
Bricklayers ______
Carpenters ______
Masons ______
Electricians ______
Iron Workers ______
Plumbers ______
Others ______
_______________ ______
Total ______

DAILY WORK LOG

7 AM ______________________________

8 AM ______________________________

9 AM ______________________________

10 AM ______________________________

11 AM ______________________________

12 NOON ______________________________

1 PM ______________________________

2 PM ______________________________

3 PM ______________________________

4 PM ______________________________

5 PM ______________________________

6 PM ______________________________

Delays / Problems ______________________________

Schedule Updates / Progress ______________________________

Extra Work / Authorized by ______________________________

Supervisor's Signature ______________________________

JOB NAME: ______________________ DATE: ____________

CONTRACTOR: ______________________ Weather: AM ______ PM ______

DAILY WORK LOG

Expenses / Materials

Material Deliveries

Equipment Use / Hours

Equipment Rentals

Daily Work Force **No.**

Daily Work Force	No.
Superintendent	
Bricklayers	
Carpenters	
Masons	
Electricians	
Iron Workers	
Plumbers	
Others	
Total	

7 AM

8 AM

9 AM

10 AM

11 AM

12 NOON

1 PM

2 PM

3 PM

4 PM

5 PM

6 PM

Delays / Problems

Schedule Updates / Progress

Extra Work / Authorized by

Supervisor's Signature ______________________

JOB NAME: ______________________ DATE: ______________

CONTRACTOR: ______________________ Weather: AM ______ PM ______

DAILY WORK LOG

Expenses / Materials

____

____

____

____

____

Material Deliveries

____

____

____

____

____

Equipment Use / Hours

____

____

____

____

____

Equipment Rentals

____

____

____

____

____

Daily Work Force	**No.**
Superintendent	______
Bricklayers	______
Carpenters	______
Masons	______
Electricians	______
Iron Workers	______
Plumbers	______
Others	______
______________	______
Total	______

7 AM ____

____

8 AM ____

____

9 AM ____

____

10 AM ____

____

11 AM ____

____

12 NOON ____

____

1 PM ____

____

2 PM ____

____

3 PM ____

____

4 PM ____

____

5 PM ____

____

6 PM ____

____

Delays / Problems ____

____

____

Schedule Updates / Progress ____

____

____

Extra Work / Authorized by ____

____

Supervisor's Signature ____

JOB NAME: ______________________ DATE: ______________

CONTRACTOR: ______________________ Weather: AM ________ PM ________

Expenses / Materials	DAILY WORK LOG
	7 AM
	8 AM
Material Deliveries	9 AM
	10 AM
	11 AM
Equipment Use / Hours	12 NOON
	1 PM
	2 PM
Equipment Rentals	3 PM
	4 PM
	5 PM
Daily Work Force No.	
Superintendent ____	6 PM
Bricklayers ____	
Carpenters ____	Delays / Problems
Masons ____	
Electricians ____	
Iron Workers ____	Schedule Updates / Progress
Plumbers ____	
Others ____	
________ ____	Extra Work / Authorized by
Total ____	

Supervisor's Signature ______________________________

JOB NAME: ______________________ DATE: ______________

CONTRACTOR: ______________________ Weather: AM ______ PM ______

Expenses / Materials

____

____

____

____

____

Material Deliveries

____

____

____

____

____

Equipment Use / Hours

____

____

____

____

____

Equipment Rentals

____

____

____

____

____

Daily Work Force	**No.**
Superintendent	______
Bricklayers	______
Carpenters	______
Masons	______
Electricians	______
Iron Workers	______
Plumbers	______
Others	______
______________	______
Total	______

DAILY WORK LOG

7 AM ______________________________

____

8 AM ______________________________

____

9 AM ______________________________

____

10 AM ______________________________

____

11 AM ______________________________

____

12 NOON ______________________________

____

1 PM ______________________________

____

2 PM ______________________________

____

3 PM ______________________________

____

4 PM ______________________________

____

5 PM ______________________________

____

6 PM ______________________________

____

Delays / Problems ______________________________

____

____

Schedule Updates / Progress ______________________________

____

____

Extra Work / Authorized by ______________________________

____

Supervisor's Signature ______________________________

JOB NAME: ______________________ DATE: __________

CONTRACTOR: ______________________ Weather: AM ______ PM ______

DAILY WORK LOG

Expenses / Materials

Material Deliveries

Equipment Use / Hours

Equipment Rentals

Daily Work Force	**No.**
Superintendent	
Bricklayers	
Carpenters	
Masons	
Electricians	
Iron Workers	
Plumbers	
Others	
Total	

7 AM

8 AM

9 AM

10 AM

11 AM

12 NOON

1 PM

2 PM

3 PM

4 PM

5 PM

6 PM

Delays / Problems

Schedule Updates / Progress

Extra Work / Authorized by

Supervisor's Signature ______________________

JOB NAME: ______________________ DATE: ______________

CONTRACTOR: ______________________ Weather: AM ______ PM ______

Expenses / Materials

____

____

____

____

____

Material Deliveries

____

____

____

____

____

Equipment Use / Hours

____

____

____

____

____

Equipment Rentals

____

____

____

____

____

Daily Work Force	**No.**
Superintendent	______
Bricklayers	______
Carpenters	______
Masons	______
Electricians	______
Iron Workers	______
Plumbers	______
Others	______
______________	______
Total	______

DAILY WORK LOG

7 AM ______________________________

____

8 AM ______________________________

____

9 AM ______________________________

____

10 AM ______________________________

____

11 AM ______________________________

____

12 NOON ______________________________

____

1 PM ______________________________

____

2 PM ______________________________

____

3 PM ______________________________

____

4 PM ______________________________

____

5 PM ______________________________

____

6 PM ______________________________

____

Delays / Problems ______________________________

____

____

Schedule Updates / Progress ______________________________

____

____

Extra Work / Authorized by ______________________________

____

Supervisor's Signature ______________________________

JOB NAME: ______________________ DATE: ______________

CONTRACTOR: ______________________ Weather: AM ________ PM ________

Expenses / Materials

DAILY WORK LOG

7 AM

8 AM

9 AM

10 AM

11 AM

12 NOON

1 PM

2 PM

3 PM

4 PM

5 PM

6 PM

Material Deliveries

Equipment Use / Hours

Equipment Rentals

Daily Work Force	**No.**
Superintendent	
Bricklayers	
Carpenters	
Masons	
Electricians	
Iron Workers	
Plumbers	
Others	
Total	

Delays / Problems

Schedule Updates / Progress

Extra Work / Authorized by

Supervisor's Signature ______________________

JOB NAME: ______________________ DATE: ______________

CONTRACTOR: ______________________ Weather: AM ______ PM ______

DAILY WORK LOG

Expenses / Materials	
	7 AM
	8 AM
Material Deliveries	9 AM
	10 AM
	11 AM
Equipment Use / Hours	
	12 NOON
	1 PM
Equipment Rentals	2 PM
	3 PM
	4 PM
Daily Work Force — No.	5 PM
Superintendent ______	
Bricklayers ______	6 PM
Carpenters ______	
Masons ______	Delays / Problems
Electricians ______	
Iron Workers ______	Schedule Updates / Progress
Plumbers ______	
Others ______	
__________ ______	Extra Work / Authorized by
Total ______	

Supervisor's Signature ______________________

JOB NAME: ______________________ DATE: ____________

CONTRACTOR: ______________________ Weather: AM ________ PM ________

DAILY WORK LOG

Expenses / Materials

Material Deliveries

Equipment Use / Hours

Equipment Rentals

Daily Work Force **No.**

Superintendent	____
Bricklayers	____
Carpenters	____
Masons	____
Electricians	____
Iron Workers	____
Plumbers	____
Others	____
____________	____
Total	____

7 AM

8 AM

9 AM

10 AM

11 AM

12 NOON

1 PM

2 PM

3 PM

4 PM

5 PM

6 PM

Delays / Problems

Schedule Updates / Progress

Extra Work / Authorized by

Supervisor's Signature ______________________

JOB NAME: ______________________ DATE: ____________

CONTRACTOR: ______________________ Weather: AM ______ PM ______

DAILY WORK LOG

Expenses / Materials

Material Deliveries

Equipment Use / Hours

Equipment Rentals

Daily Work Force **No.**

Superintendent ______

Bricklayers ______

Carpenters ______

Masons ______

Electricians ______

Iron Workers ______

Plumbers ______

Others ______

Total ______

7 AM

8 AM

9 AM

10 AM

11 AM

12 NOON

1 PM

2 PM

3 PM

4 PM

5 PM

6 PM

Delays / Problems

Schedule Updates / Progress

Extra Work / Authorized by

Supervisor's Signature ______________________

JOB NAME: ______________________ DATE: __________

CONTRACTOR: ______________________ Weather: AM ______ PM ______

DAILY WORK LOG

Expenses / Materials

Material Deliveries

Equipment Use / Hours

Equipment Rentals

Daily Work Force **No.**

Superintendent ____

Bricklayers ____

Carpenters ____

Masons ____

Electricians ____

Iron Workers ____

Plumbers ____

Others ____

____________ ____

Total ____

7 AM

8 AM

9 AM

10 AM

11 AM

12 NOON

1 PM

2 PM

3 PM

4 PM

5 PM

6 PM

Delays / Problems

Schedule Updates / Progress

Extra Work / Authorized by

Supervisor's Signature ______________________

JOB NAME: ______________________ DATE: ______________

CONTRACTOR: ______________________ Weather: AM ______ PM ______

DAILY WORK LOG

Expenses / Materials

____

____

____

____

____

Material Deliveries

____

____

____

____

____

Equipment Use / Hours

____

____

____

____

____

Equipment Rentals

____

____

____

____

____

Daily Work Force	**No.**
Superintendent	______
Bricklayers	______
Carpenters	______
Masons	______
Electricians	______
Iron Workers	______
Plumbers	______
Others	______
______________	______
Total	______

7 AM ______________________________________

8 AM ______________________________________

9 AM ______________________________________

10 AM ______________________________________

11 AM ______________________________________

12 NOON ______________________________________

1 PM ______________________________________

2 PM ______________________________________

3 PM ______________________________________

4 PM ______________________________________

5 PM ______________________________________

6 PM ______________________________________

Delays / Problems ______________________________________

Schedule Updates / Progress ______________________________________

Extra Work / Authorized by ______________________________________

Supervisor's Signature ______________________________________

JOB NAME: ______________________ DATE: ______________

CONTRACTOR: ______________________ Weather: AM ______ PM ______

DAILY WORK LOG

Expenses / Materials

____

Material Deliveries

____

Equipment Use / Hours

____

Equipment Rentals

____

Daily Work Force	**No.**
Superintendent	____
Bricklayers	____
Carpenters	____
Masons	____
Electricians	____
Iron Workers	____
Plumbers	____
Others	____
____________	____
Total	____

7 AM ______________________

8 AM ______________________

9 AM ______________________

10 AM ______________________

11 AM ______________________

12 NOON ______________________

1 PM ______________________

2 PM ______________________

3 PM ______________________

4 PM ______________________

5 PM ______________________

6 PM ______________________

Delays / Problems ______________________

Schedule Updates / Progress ______________________

Extra Work / Authorized by ______________________

Supervisor's Signature ______________________

JOB NAME: ______________________ DATE: ______________

CONTRACTOR: ______________________ Weather: AM ______ PM ______

Expenses / Materials

____

____

____

____

____

Material Deliveries

____

____

____

____

____

Equipment Use / Hours

____

____

____

____

____

Equipment Rentals

____

____

____

____

____

Daily Work Force	**No.**
Superintendent	______
Bricklayers	______
Carpenters	______
Masons	______
Electricians	______
Iron Workers	______
Plumbers	______
Others	______
______________	______
Total	______

DAILY WORK LOG

7 AM ______________________________

____

8 AM ______________________________

____

9 AM ______________________________

____

10 AM ______________________________

____

11 AM ______________________________

____

12 NOON ______________________________

____

1 PM ______________________________

____

2 PM ______________________________

____

3 PM ______________________________

____

4 PM ______________________________

____

5 PM ______________________________

____

6 PM ______________________________

____

Delays / Problems ______________________________

____

____

Schedule Updates / Progress ______________________________

____

____

Extra Work / Authorized by ______________________________

____

Supervisor's Signature ______________________________

JOB NAME: ______________________ DATE: ______________

CONTRACTOR: ______________________ Weather: AM ______ PM ______

DAILY WORK LOG

Expenses / Materials

Material Deliveries

Equipment Use / Hours

Equipment Rentals

Daily Work Force **No.**

Superintendent ______

Bricklayers ______

Carpenters ______

Masons ______

Electricians ______

Iron Workers ______

Plumbers ______

Others ______

______________ ______

Total ______

7 AM

8 AM

9 AM

10 AM

11 AM

12 NOON

1 PM

2 PM

3 PM

4 PM

5 PM

6 PM

Delays / Problems

Schedule Updates / Progress

Extra Work / Authorized by

Supervisor's Signature ______________________

JOB NAME: ______________________ DATE: ______________

CONTRACTOR: ______________________ Weather: AM ______ PM ______

DAILY WORK LOG

Expenses / Materials

____

____

____

____

____

Material Deliveries

____

____

____

____

____

Equipment Use / Hours

____

____

____

____

____

Equipment Rentals

____

____

____

____

____

Daily Work Force	**No.**
Superintendent	______
Bricklayers	______
Carpenters	______
Masons	______
Electricians	______
Iron Workers	______
Plumbers	______
Others	______
______________	______
Total	______

7 AM ______________________________________

____

8 AM ______________________________________

____

9 AM ______________________________________

____

10 AM ______________________________________

____

11 AM ______________________________________

____

12 NOON ______________________________________

____

1 PM ______________________________________

____

2 PM ______________________________________

____

3 PM ______________________________________

____

4 PM ______________________________________

____

5 PM ______________________________________

____

6 PM ______________________________________

____

Delays / Problems ______________________________________

____

____

Schedule Updates / Progress ______________________________________

____

____

Extra Work / Authorized by ______________________________________

____

Supervisor's Signature ______________________________________

JOB NAME: ______________________ DATE: ______________

CONTRACTOR: ______________________ Weather: AM ________ PM ________

Expenses / Materials	DAILY WORK LOG
	7 AM
	8 AM
Material Deliveries	9 AM
	10 AM
	11 AM
Equipment Use / Hours	12 NOON
	1 PM
	2 PM
Equipment Rentals	3 PM
	4 PM
	5 PM
Daily Work Force **No.**	6 PM
Superintendent ____	
Bricklayers ____	
Carpenters ____	Delays / Problems
Masons ____	
Electricians ____	
Iron Workers ____	Schedule Updates / Progress
Plumbers ____	
Others ____	
____________ ____	Extra Work / Authorized by
Total ____	

Supervisor's Signature ______________________________

JOB NAME: ______________________ DATE: ______________

CONTRACTOR: ______________________ Weather: AM ______ PM ______

DAILY WORK LOG

Expenses / Materials	

Material Deliveries

_____
_____
_____
_____
_____

Equipment Use / Hours

_____
_____
_____
_____
_____

Equipment Rentals

_____
_____
_____
_____
_____

Daily Work Force	**No.**
Superintendent	
Bricklayers	
Carpenters	
Masons	
Electricians	
Iron Workers	
Plumbers	
Others	
Total	

7 AM

8 AM

9 AM

10 AM

11 AM

12 NOON

1 PM

2 PM

3 PM

4 PM

5 PM

6 PM

Delays / Problems

Schedule Updates / Progress

Extra Work / Authorized by

Supervisor's Signature ______________________

JOB NAME: ______________________ DATE: __________

CONTRACTOR: ______________________ Weather: AM ______ PM ______

DAILY WORK LOG

Expenses / Materials

Material Deliveries

Equipment Use / Hours

Equipment Rentals

Daily Work Force	**No.**
Superintendent	
Bricklayers	
Carpenters	
Masons	
Electricians	
Iron Workers	
Plumbers	
Others	
Total	

7 AM

8 AM

9 AM

10 AM

11 AM

12 NOON

1 PM

2 PM

3 PM

4 PM

5 PM

6 PM

Delays / Problems

Schedule Updates / Progress

Extra Work / Authorized by

Supervisor's Signature ______________________

JOB NAME: ______________________ DATE: ______________

CONTRACTOR: ______________________ Weather: AM ______ PM ______

DAILY WORK LOG

Expenses / Materials	
	7 AM
	8 AM
Material Deliveries	9 AM
	10 AM
	11 AM
Equipment Use / Hours	12 NOON
	1 PM
	2 PM
Equipment Rentals	3 PM
	4 PM
	5 PM
Daily Work Force **No.**	6 PM
Superintendent ______	
Bricklayers ______	
Carpenters ______	Delays / Problems
Masons ______	
Electricians ______	
Iron Workers ______	Schedule Updates / Progress
Plumbers ______	
Others ______	
______ ______	Extra Work / Authorized by
Total ______	

Supervisor's Signature ______________________

JOB NAME: ______________________ DATE: ______________

CONTRACTOR: ______________________ Weather: AM ________ PM ________

DAILY WORK LOG

Expenses / Materials	
Material Deliveries	
Equipment Use / Hours	
Equipment Rentals	

Daily Work Force **No.**

	No.
Superintendent	
Bricklayers	
Carpenters	
Masons	
Electricians	
Iron Workers	
Plumbers	
Others	
Total	

7 AM

8 AM

9 AM

10 AM

11 AM

12 NOON

1 PM

2 PM

3 PM

4 PM

5 PM

6 PM

Delays / Problems

Schedule Updates / Progress

Extra Work / Authorized by

Supervisor's Signature ______________________

JOB NAME: ______________________ DATE: ______________

CONTRACTOR: ______________________ Weather: AM ______ PM ______

Expenses / Materials	DAILY WORK LOG
	7 AM
	8 AM
Material Deliveries	9 AM
	10 AM
	11 AM
Equipment Use / Hours	12 NOON
	1 PM
	2 PM
Equipment Rentals	3 PM
	4 PM
	5 PM
Daily Work Force **No.**	6 PM
Superintendent ______	
Bricklayers ______	
Carpenters ______	Delays / Problems
Masons ______	
Electricians ______	
Iron Workers ______	Schedule Updates / Progress
Plumbers ______	
Others ______	
____________ ______	Extra Work / Authorized by
Total ______	

Supervisor's Signature ______________________________

JOB NAME: ______________________ DATE: ______________

CONTRACTOR: ______________________ Weather: AM ______ PM ______

DAILY WORK LOG

Expenses / Materials

Material Deliveries

Equipment Use / Hours

Equipment Rentals

Daily Work Force **No.**

Superintendent	____
Bricklayers	____
Carpenters	____
Masons	____
Electricians	____
Iron Workers	____
Plumbers	____
Others	____
____________	____
Total	____

7 AM

8 AM

9 AM

10 AM

11 AM

12 NOON

1 PM

2 PM

3 PM

4 PM

5 PM

6 PM

Delays / Problems

Schedule Updates / Progress

Extra Work / Authorized by

Supervisor's Signature ______________________

JOB NAME: ______________________________ DATE: ______________

CONTRACTOR: ______________________________ Weather: AM ______ PM ______

DAILY WORK LOG

Expenses / Materials

____

____

____

____

____

Material Deliveries

____

____

____

____

____

Equipment Use / Hours

____

____

____

____

____

Equipment Rentals

____

____

____

____

____

Daily Work Force	**No.**
Superintendent	______
Bricklayers	______
Carpenters	______
Masons	______
Electricians	______
Iron Workers	______
Plumbers	______
Others	______
______________	______
Total	______

7 AM ______________________________

8 AM ______________________________

9 AM ______________________________

10 AM ______________________________

11 AM ______________________________

12 NOON ______________________________

1 PM ______________________________

2 PM ______________________________

3 PM ______________________________

4 PM ______________________________

5 PM ______________________________

6 PM ______________________________

Delays / Problems ______________________________

Schedule Updates / Progress ______________________________

Extra Work / Authorized by ______________________________

Supervisor's Signature ______________________________

JOB NAME: ______________________ DATE: ______________

CONTRACTOR: ______________________ Weather: AM ________ PM ________

DAILY WORK LOG

Expenses / Materials

____

____

____

____

Material Deliveries

____

____

____

____

Equipment Use / Hours

____

____

____

____

Equipment Rentals

____

____

____

____

Daily Work Force	**No.**
Superintendent	______
Bricklayers	______
Carpenters	______
Masons	______
Electricians	______
Iron Workers	______
Plumbers	______
Others	______
____________	______
Total	______

7 AM ______________________

8 AM ______________________

9 AM ______________________

10 AM ______________________

11 AM ______________________

12 NOON ______________________

1 PM ______________________

2 PM ______________________

3 PM ______________________

4 PM ______________________

5 PM ______________________

6 PM ______________________

Delays / Problems ______________________

Schedule Updates / Progress ______________________

Extra Work / Authorized by ______________________

Supervisor's Signature ______________________

JOB NAME: ______________________ DATE: ______________

CONTRACTOR: ______________________ Weather: AM ______ PM ______

Expenses / Materials

Material Deliveries

Equipment Use / Hours

Equipment Rentals

Daily Work Force **No.**

Superintendent	____
Bricklayers	____
Carpenters	____
Masons	____
Electricians	____
Iron Workers	____
Plumbers	____
Others	____
________	____
Total	____

DAILY WORK LOG

7 AM

8 AM

9 AM

10 AM

11 AM

12 NOON

1 PM

2 PM

3 PM

4 PM

5 PM

6 PM

Delays / Problems

Schedule Updates / Progress

Extra Work / Authorized by

Supervisor's Signature ______________________

JOB NAME: ______________________ DATE: ____________

CONTRACTOR: ______________________ Weather: AM ________ PM ________

DAILY WORK LOG

Expenses / Materials

Material Deliveries

Equipment Use / Hours

Equipment Rentals

Daily Work Force **No.**

Superintendent	______
Bricklayers	______
Carpenters	______
Masons	______
Electricians	______
Iron Workers	______
Plumbers	______
Others	______
____________	______
Total	______

7 AM

8 AM

9 AM

10 AM

11 AM

12 NOON

1 PM

2 PM

3 PM

4 PM

5 PM

6 PM

Delays / Problems

Schedule Updates / Progress

Extra Work / Authorized by

Supervisor's Signature ______________________

JOB NAME: ____________________ DATE: __________

CONTRACTOR: ____________________ Weather: AM ______ PM ______

DAILY WORK LOG

Expenses / Materials	
	7 AM
	8 AM
Material Deliveries	9 AM
	10 AM
	11 AM
Equipment Use / Hours	12 NOON
	1 PM
	2 PM
Equipment Rentals	
	3 PM
	4 PM
	5 PM
	6 PM

Daily Work Force	**No.**
Superintendent	
Bricklayers	
Carpenters	
Masons	
Electricians	
Iron Workers	
Plumbers	
Others	
Total	

Delays / Problems

Schedule Updates / Progress

Extra Work / Authorized by

Supervisor's Signature ____________________

JOB NAME: ____________________ DATE: ____________

CONTRACTOR: ____________________ Weather: AM ______ PM ______

DAILY WORK LOG

Expenses / Materials

Material Deliveries

Equipment Use / Hours

Equipment Rentals

Daily Work Force **No.**

Superintendent	
Bricklayers	
Carpenters	
Masons	
Electricians	
Iron Workers	
Plumbers	
Others	
Total	

7 AM

8 AM

9 AM

10 AM

11 AM

12 NOON

1 PM

2 PM

3 PM

4 PM

5 PM

6 PM

Delays / Problems

Schedule Updates / Progress

Extra Work / Authorized by

Supervisor's Signature ____________________

JOB NAME: ______________________ DATE: ____________

CONTRACTOR: ______________________ Weather: AM ______ PM ______

Expenses / Materials

Material Deliveries

Equipment Use / Hours

Equipment Rentals

Daily Work Force **No.**

Superintendent	____
Bricklayers	____
Carpenters	____
Masons	____
Electricians	____
Iron Workers	____
Plumbers	____
Others	____
________	____
Total	____

DAILY WORK LOG

7 AM

8 AM

9 AM

10 AM

11 AM

12 NOON

1 PM

2 PM

3 PM

4 PM

5 PM

6 PM

Delays / Problems

Schedule Updates / Progress

Extra Work / Authorized by

Supervisor's Signature ______________________

JOB NAME: ______________________ DATE: ____________

CONTRACTOR: ______________________ Weather: AM ______ PM ______

Expenses / Materials

Material Deliveries

Equipment Use / Hours

Equipment Rentals

Daily Work Force No.

Superintendent ______

Bricklayers ______

Carpenters ______

Masons ______

Electricians ______

Iron Workers ______

Plumbers ______

Others ______

______________ ______

Total ______

DAILY WORK LOG

7 AM

8 AM

9 AM

10 AM

11 AM

12 NOON

1 PM

2 PM

3 PM

4 PM

5 PM

6 PM

Delays / Problems

Schedule Updates / Progress

Extra Work / Authorized by

Supervisor's Signature ______________________

JOB NAME: ______________________ DATE: ______________

CONTRACTOR: ______________________ Weather: AM ______ PM ______

Expenses / Materials

Material Deliveries

Equipment Use / Hours

Equipment Rentals

Daily Work Force **No.**

Superintendent	______
Bricklayers	______
Carpenters	______
Masons	______
Electricians	______
Iron Workers	______
Plumbers	______
Others	______
______________	______
Total	______

DAILY WORK LOG

7 AM

8 AM

9 AM

10 AM

11 AM

12 NOON

1 PM

2 PM

3 PM

4 PM

5 PM

6 PM

Delays / Problems

Schedule Updates / Progress

Extra Work / Authorized by

Supervisor's Signature ______________________________

JOB NAME: ______________________ DATE: ____________

CONTRACTOR: ______________________ Weather: AM ______ PM ______

DAILY WORK LOG

Expenses / Materials

____

____

____

____

Material Deliveries

____

____

____

____

Equipment Use / Hours

____

____

____

____

Equipment Rentals

____

____

____

____

Daily Work Force	**No.**
Superintendent	______
Bricklayers	______
Carpenters	______
Masons	______
Electricians	______
Iron Workers	______
Plumbers	______
Others	______
______________	______
Total	______

7 AM ______________________

8 AM ______________________

9 AM ______________________

10 AM ______________________

11 AM ______________________

12 NOON ______________________

1 PM ______________________

2 PM ______________________

3 PM ______________________

4 PM ______________________

5 PM ______________________

6 PM ______________________

Delays / Problems ______________________

Schedule Updates / Progress ______________________

Extra Work / Authorized by ______________________

Supervisor's Signature ______________________

JOB NAME: ______________________ DATE: ______________

CONTRACTOR: ______________________ Weather: AM ______ PM ______

Expenses / Materials

____
____
____
____
____

Material Deliveries

____
____
____
____
____

Equipment Use / Hours

____
____
____
____
____

Equipment Rentals

____
____
____
____
____

Daily Work Force	**No.**
Superintendent	______
Bricklayers	______
Carpenters	______
Masons	______
Electricians	______
Iron Workers	______
Plumbers	______
Others	______
______________	______
Total	______

DAILY WORK LOG

7 AM ______________________________________
____
8 AM ______________________________________
____
9 AM ______________________________________
____
10 AM ______________________________________
____
11 AM ______________________________________
____
12 NOON ______________________________________
____
1 PM ______________________________________
____
2 PM ______________________________________
____
3 PM ______________________________________
____
4 PM ______________________________________
____
5 PM ______________________________________
____
6 PM ______________________________________
____

Delays / Problems ______________________________________
____
____

Schedule Updates / Progress ______________________________________
____
____

Extra Work / Authorized by ______________________________________
____

Supervisor's Signature ______________________________________

JOB NAME: ______________________ DATE: ____________

CONTRACTOR: ______________________ Weather: AM ______ PM ______

DAILY WORK LOG

Expenses / Materials

Material Deliveries

Equipment Use / Hours

Equipment Rentals

Daily Work Force **No.**

Daily Work Force	No.
Superintendent	
Bricklayers	
Carpenters	
Masons	
Electricians	
Iron Workers	
Plumbers	
Others	
Total	

7 AM

8 AM

9 AM

10 AM

11 AM

12 NOON

1 PM

2 PM

3 PM

4 PM

5 PM

6 PM

Delays / Problems

Schedule Updates / Progress

Extra Work / Authorized by

Supervisor's Signature ______________________

JOB NAME: ____________________ DATE: ____________

CONTRACTOR: ____________________ Weather: AM ______ PM ______

Expenses / Materials

Material Deliveries

Equipment Use / Hours

Equipment Rentals

Daily Work Force **No.**

Superintendent ______

Bricklayers ______

Carpenters ______

Masons ______

Electricians ______

Iron Workers ______

Plumbers ______

Others ______

______ ______

Total ______

DAILY WORK LOG

7 AM

8 AM

9 AM

10 AM

11 AM

12 NOON

1 PM

2 PM

3 PM

4 PM

5 PM

6 PM

Delays / Problems

Schedule Updates / Progress

Extra Work / Authorized by

Supervisor's Signature ____________________

JOB NAME: ______________________ DATE: __________

CONTRACTOR: ______________________ Weather: AM ______ PM ______

DAILY WORK LOG

Expenses / Materials

Material Deliveries

Equipment Use / Hours

Equipment Rentals

Daily Work Force	**No.**
Superintendent	
Bricklayers	
Carpenters	
Masons	
Electricians	
Iron Workers	
Plumbers	
Others	
Total	

7 AM

8 AM

9 AM

10 AM

11 AM

12 NOON

1 PM

2 PM

3 PM

4 PM

5 PM

6 PM

Delays / Problems

Schedule Updates / Progress

Extra Work / Authorized by

Supervisor's Signature ______________________

JOB NAME: ______________________ DATE: ____________

CONTRACTOR: ______________________ Weather: AM ______ PM ______

DAILY WORK LOG

Expenses / Materials

____

____

____

____

____

Material Deliveries

____

____

____

____

____

Equipment Use / Hours

____

____

____

____

____

Equipment Rentals

____

____

____

____

____

Daily Work Force	**No.**
Superintendent	______
Bricklayers	______
Carpenters	______
Masons	______
Electricians	______
Iron Workers	______
Plumbers	______
Others	______
______________	______
Total	______

7 AM ______________________________

____

8 AM ______________________________

____

9 AM ______________________________

____

10 AM ______________________________

____

11 AM ______________________________

____

12 NOON ______________________________

____

1 PM ______________________________

____

2 PM ______________________________

____

3 PM ______________________________

____

4 PM ______________________________

____

5 PM ______________________________

____

6 PM ______________________________

____

Delays / Problems ______________________________

____

____

Schedule Updates / Progress ______________________________

____

____

Extra Work / Authorized by ______________________________

____

Supervisor's Signature ______________________________

JOB NAME: ______________________ DATE: ____________

CONTRACTOR: ______________________ Weather: AM ______ PM ______

DAILY WORK LOG

Expenses / Materials	7 AM
	8 AM
Material Deliveries	9 AM
	10 AM
	11 AM
Equipment Use / Hours	12 NOON
	1 PM
	2 PM
Equipment Rentals	3 PM
	4 PM
	5 PM
Daily Work Force **No.**	6 PM
Superintendent ______	
Bricklayers ______	
Carpenters ______	Delays / Problems
Masons ______	
Electricians ______	
Iron Workers ______	Schedule Updates / Progress
Plumbers ______	
Others ______	
____________ ______	Extra Work / Authorized by
Total ______	

Supervisor's Signature ______________________

JOB NAME: ______________________ DATE: ______________

CONTRACTOR: ______________________ Weather: AM ______ PM ______

DAILY WORK LOG

Expenses / Materials

____

____

____

____

____

Material Deliveries

____

____

____

____

____

Equipment Use / Hours

____

____

____

____

____

Equipment Rentals

____

____

____

____

____

Daily Work Force	**No.**
Superintendent	______
Bricklayers	______
Carpenters	______
Masons	______
Electricians	______
Iron Workers	______
Plumbers	______
Others	______
______________	______
Total	______

7 AM ______________________________

____

8 AM ______________________________

____

9 AM ______________________________

____

10 AM ______________________________

____

11 AM ______________________________

____

12 NOON ______________________________

____

1 PM ______________________________

____

2 PM ______________________________

____

3 PM ______________________________

____

4 PM ______________________________

____

5 PM ______________________________

____

6 PM ______________________________

____

Delays / Problems ______________________________

____

____

Schedule Updates / Progress ______________________________

____

____

Extra Work / Authorized by ______________________________

____

Supervisor's Signature ______________________________

JOB NAME: ______________________ DATE: __________

CONTRACTOR: ______________________ Weather: AM ______ PM ______

DAILY WORK LOG

Expenses / Materials

Material Deliveries

Equipment Use / Hours

Equipment Rentals

Daily Work Force **No.**

Superintendent ______

Bricklayers ______

Carpenters ______

Masons ______

Electricians ______

Iron Workers ______

Plumbers ______

Others ______

__________ ______

Total ______

7 AM

8 AM

9 AM

10 AM

11 AM

12 NOON

1 PM

2 PM

3 PM

4 PM

5 PM

6 PM

Delays / Problems

Schedule Updates / Progress

Extra Work / Authorized by

Supervisor's Signature ______________________

JOB NAME: ______________________ DATE: ____________

CONTRACTOR: ______________________ Weather: AM ______ PM ______

Expenses / Materials

____

____

____

____

____

Material Deliveries

____

____

____

____

____

Equipment Use / Hours

____

____

____

____

____

Equipment Rentals

____

____

____

____

____

Daily Work Force	**No.**
Superintendent	______
Bricklayers	______
Carpenters	______
Masons	______
Electricians	______
Iron Workers	______
Plumbers	______
Others	______
____________	______
Total	______

DAILY WORK LOG

7 AM ______________________________

____

8 AM ______________________________

____

9 AM ______________________________

____

10 AM ______________________________

____

11 AM ______________________________

____

12 NOON ______________________________

____

1 PM ______________________________

____

2 PM ______________________________

____

3 PM ______________________________

____

4 PM ______________________________

____

5 PM ______________________________

____

6 PM ______________________________

____

Delays / Problems ______________________________

____

____

Schedule Updates / Progress ______________________________

____

____

Extra Work / Authorized by ______________________________

____

Supervisor's Signature ______________________________

JOB NAME: ______________________ DATE: ______________

CONTRACTOR: ______________________ Weather: AM ______ PM ______

DAILY WORK LOG

Expenses / Materials

Material Deliveries

Equipment Use / Hours

Equipment Rentals

Daily Work Force **No.**

Superintendent ______

Bricklayers ______

Carpenters ______

Masons ______

Electricians ______

Iron Workers ______

Plumbers ______

Others ______

______________ ______

Total ______

7 AM

8 AM

9 AM

10 AM

11 AM

12 NOON

1 PM

2 PM

3 PM

4 PM

5 PM

6 PM

Delays / Problems

Schedule Updates / Progress

Extra Work / Authorized by

Supervisor's Signature ______________________

JOB NAME: ______________________ DATE: ______________

CONTRACTOR: ______________________ Weather: AM ______ PM ______

DAILY WORK LOG

Expenses / Materials

____
____
____
____
____

Material Deliveries

____
____
____
____
____

Equipment Use / Hours

____
____
____
____
____

Equipment Rentals

____
____
____
____
____

Daily Work Force	**No.**
Superintendent	______
Bricklayers	______
Carpenters	______
Masons	______
Electricians	______
Iron Workers	______
Plumbers	______
Others	______
______________	______
Total	______

7 AM ______________________________

8 AM ______________________________

9 AM ______________________________

10 AM ______________________________

11 AM ______________________________

12 NOON ______________________________

1 PM ______________________________

2 PM ______________________________

3 PM ______________________________

4 PM ______________________________

5 PM ______________________________

6 PM ______________________________

Delays / Problems ______________________________

Schedule Updates / Progress ______________________________

Extra Work / Authorized by ______________________________

Supervisor's Signature ______________________________

JOB NAME: ____________________ DATE: ____________

CONTRACTOR: ____________________ Weather: AM ______ PM ______

DAILY WORK LOG

Expenses / Materials

Material Deliveries

Equipment Use / Hours

Equipment Rentals

Daily Work Force **No.**

Superintendent ______
Bricklayers ______
Carpenters ______
Masons ______
Electricians ______
Iron Workers ______
Plumbers ______
Others ______
____________ ______

Total ______

7 AM

8 AM

9 AM

10 AM

11 AM

12 NOON

1 PM

2 PM

3 PM

4 PM

5 PM

6 PM

Delays / Problems

Schedule Updates / Progress

Extra Work / Authorized by

Supervisor's Signature ______________________________

JOB NAME: ______________________ DATE: ______________

CONTRACTOR: ______________________ Weather: AM ______ PM ______

DAILY WORK LOG

Expenses / Materials

____

____

____

____

____

Material Deliveries

____

____

____

____

____

Equipment Use / Hours

____

____

____

____

____

Equipment Rentals

____

____

____

____

____

Daily Work Force	**No.**
Superintendent	______
Bricklayers	______
Carpenters	______
Masons	______
Electricians	______
Iron Workers	______
Plumbers	______
Others	______
______________	______
Total	______

7 AM ______________________

____

8 AM ______________________

____

9 AM ______________________

____

10 AM ______________________

____

11 AM ______________________

____

12 NOON ______________________

____

1 PM ______________________

____

2 PM ______________________

____

3 PM ______________________

____

4 PM ______________________

____

5 PM ______________________

____

6 PM ______________________

____

Delays / Problems ______________________

____

____

Schedule Updates / Progress ______________________

____

____

Extra Work / Authorized by ______________________

____

Supervisor's Signature ______________________

JOB NAME: ______________________ DATE: ______________

CONTRACTOR: ______________________ Weather: AM ______ PM ______

DAILY WORK LOG

Expenses / Materials

____

____

____

____

____

Material Deliveries

____

____

____

____

____

Equipment Use / Hours

____

____

____

____

____

Equipment Rentals

____

____

____

____

____

Daily Work Force	**No.**
Superintendent	______
Bricklayers	______
Carpenters	______
Masons	______
Electricians	______
Iron Workers	______
Plumbers	______
Others	______
______________	______
Total	______

7 AM ______________________________

____

8 AM ______________________________

____

9 AM ______________________________

____

10 AM ______________________________

____

11 AM ______________________________

____

12 NOON ______________________________

____

1 PM ______________________________

____

2 PM ______________________________

____

3 PM ______________________________

____

4 PM ______________________________

____

5 PM ______________________________

____

6 PM ______________________________

____

Delays / Problems ______________________________

____

____

Schedule Updates / Progress ______________________________

____

____

Extra Work / Authorized by ______________________________

____

Supervisor's Signature ______________________________

JOB NAME: ______________________ DATE: ____________

CONTRACTOR: ______________________ Weather: AM ______ PM ______

Expenses / Materials

____

____

____

____

____

Material Deliveries

____

____

____

____

____

Equipment Use / Hours

____

____

____

____

____

Equipment Rentals

____

____

____

____

____

Daily Work Force	**No.**
Superintendent	______
Bricklayers	______
Carpenters	______
Masons	______
Electricians	______
Iron Workers	______
Plumbers	______
Others	______
____________	______
Total	______

DAILY WORK LOG

7 AM ______________________

____

8 AM ______________________

____

9 AM ______________________

____

10 AM ______________________

____

11 AM ______________________

____

12 NOON ______________________

____

1 PM ______________________

____

2 PM ______________________

____

3 PM ______________________

____

4 PM ______________________

____

5 PM ______________________

____

6 PM ______________________

____

Delays / Problems ______________________

____

____

Schedule Updates / Progress ______________________

____

____

Extra Work / Authorized by ______________________

____

Supervisor's Signature ______________________

JOB NAME: ______________________ DATE: ______________

CONTRACTOR: ______________________ Weather: AM ______ PM ______

DAILY WORK LOG

Expenses / Materials

Material Deliveries

Equipment Use / Hours

Equipment Rentals

Daily Work Force **No.**

Superintendent ____

Bricklayers ____

Carpenters ____

Masons ____

Electricians ____

Iron Workers ____

Plumbers ____

Others ____

________ ____

Total ____

7 AM

8 AM

9 AM

10 AM

11 AM

12 NOON

1 PM

2 PM

3 PM

4 PM

5 PM

6 PM

Delays / Problems

Schedule Updates / Progress

Extra Work / Authorized by

Supervisor's Signature ______________________

JOB NAME: ______________________ DATE: ______________

CONTRACTOR: ______________________ Weather: AM ______ PM ______

DAILY WORK LOG

Expenses / Materials

Material Deliveries

Equipment Use / Hours

Equipment Rentals

7 AM

8 AM

9 AM

10 AM

11 AM

12 NOON

1 PM

2 PM

3 PM

4 PM

5 PM

6 PM

Daily Work Force **No.**

Daily Work Force	No.
Superintendent	
Bricklayers	
Carpenters	
Masons	
Electricians	
Iron Workers	
Plumbers	
Others	
Total	

Delays / Problems

Schedule Updates / Progress

Extra Work / Authorized by

Supervisor's Signature ______________________

JOB NAME: ______________________ DATE: ______________

CONTRACTOR: ______________________ Weather: AM ______ PM ______

DAILY WORK LOG

Expenses / Materials

Material Deliveries

Equipment Use / Hours

Equipment Rentals

Daily Work Force **No.**

Superintendent	______
Bricklayers	______
Carpenters	______
Masons	______
Electricians	______
Iron Workers	______
Plumbers	______
Others	______
____________	______
Total	______

7 AM

8 AM

9 AM

10 AM

11 AM

12 NOON

1 PM

2 PM

3 PM

4 PM

5 PM

6 PM

Delays / Problems

Schedule Updates / Progress

Extra Work / Authorized by

Supervisor's Signature ______________________

JOB NAME: ______________________ DATE: ______________

CONTRACTOR: ______________________ Weather: AM ______ PM ______

Expenses / Materials

____

____

____

____

____

Material Deliveries

____

____

____

____

____

Equipment Use / Hours

____

____

____

____

____

Equipment Rentals

____

____

____

____

____

Daily Work Force	**No.**
Superintendent	______
Bricklayers	______
Carpenters	______
Masons	______
Electricians	______
Iron Workers	______
Plumbers	______
Others	______
______________	______
Total	______

DAILY WORK LOG

7 AM ______________________________

____

8 AM ______________________________

____

9 AM ______________________________

____

10 AM ______________________________

____

11 AM ______________________________

____

12 NOON ______________________________

____

1 PM ______________________________

____

2 PM ______________________________

____

3 PM ______________________________

____

4 PM ______________________________

____

5 PM ______________________________

____

6 PM ______________________________

____

Delays / Problems ______________________________

____

____

Schedule Updates / Progress ______________________________

____

____

Extra Work / Authorized by ______________________________

____

Supervisor's Signature ______________________________

JOB NAME: ______________________ DATE: __________

CONTRACTOR: ______________________ Weather: AM ______ PM ______

DAILY WORK LOG

Expenses / Materials	
	7 AM
	8 AM
Material Deliveries	9 AM
	10 AM
	11 AM
Equipment Use / Hours	12 NOON
	1 PM
	2 PM
Equipment Rentals	3 PM
	4 PM
	5 PM

Daily Work Force **No.**

Superintendent ______

Bricklayers ______

Carpenters ______

Masons ______

Electricians ______

Iron Workers ______

Plumbers ______

Others ______

__________ ______

Total ______

6 PM

Delays / Problems

Schedule Updates / Progress

Extra Work / Authorized by

Supervisor's Signature ______________________

JOB NAME: ______________________ DATE: ______________

CONTRACTOR: ______________________ Weather: AM ______ PM ______

DAILY WORK LOG

Expenses / Materials

______
______
______
______
______

Material Deliveries

______
______
______
______
______

Equipment Use / Hours

______
______
______
______
______

Equipment Rentals

______
______
______
______
______

Daily Work Force	**No.**
Superintendent	______
Bricklayers	______
Carpenters	______
Masons	______
Electricians	______
Iron Workers	______
Plumbers	______
Others	______
______	______
Total	______

7 AM ______
______

8 AM ______
______

9 AM ______
______

10 AM ______
______

11 AM ______
______

12 NOON ______
______

1 PM ______
______

2 PM ______
______

3 PM ______
______

4 PM ______
______

5 PM ______
______

6 PM ______
______

Delays / Problems ______
______
______

Schedule Updates / Progress ______
______
______

Extra Work / Authorized by ______
______

Supervisor's Signature ______________________

JOB NAME: ______________________ DATE: ______________

CONTRACTOR: ______________________ Weather: AM ________ PM ________

DAILY WORK LOG

Expenses / Materials

Material Deliveries

Equipment Use / Hours

Equipment Rentals

Daily Work Force **No.**

Superintendent ______

Bricklayers ______

Carpenters ______

Masons ______

Electricians ______

Iron Workers ______

Plumbers ______

Others ______

______________ ______

Total ______

7 AM

8 AM

9 AM

10 AM

11 AM

12 NOON

1 PM

2 PM

3 PM

4 PM

5 PM

6 PM

Delays / Problems

Schedule Updates / Progress

Extra Work / Authorized by

Supervisor's Signature ______________________

JOB NAME: ______________________ DATE: ______________

CONTRACTOR: ______________________ Weather: AM ______ PM ______

DAILY WORK LOG

Expenses / Materials

____

____

____

____

____

Material Deliveries

____

____

____

____

____

Equipment Use / Hours

____

____

____

____

____

Equipment Rentals

____

____

____

____

____

Daily Work Force	**No.**
Superintendent	______
Bricklayers	______
Carpenters	______
Masons	______
Electricians	______
Iron Workers	______
Plumbers	______
Others	______
______________	______
Total	______

7 AM ______________________________________

____

8 AM ______________________________________

____

9 AM ______________________________________

____

10 AM ______________________________________

____

11 AM ______________________________________

____

12 NOON ______________________________________

____

1 PM ______________________________________

____

2 PM ______________________________________

____

3 PM ______________________________________

____

4 PM ______________________________________

____

5 PM ______________________________________

____

6 PM ______________________________________

____

Delays / Problems ______________________________________

____

____

Schedule Updates / Progress ______________________________________

____

____

Extra Work / Authorized by ______________________________________

____

Supervisor's Signature ______________________________________

JOB NAME: ______________________ DATE: ______________

CONTRACTOR: ______________________ Weather: AM ______ PM ______

DAILY WORK LOG

Expenses / Materials

Material Deliveries

Equipment Use / Hours

Equipment Rentals

Daily Work Force **No.**

Superintendent ______

Bricklayers ______

Carpenters ______

Masons ______

Electricians ______

Iron Workers ______

Plumbers ______

Others ______

______ ______

Total ______

7 AM

8 AM

9 AM

10 AM

11 AM

12 NOON

1 PM

2 PM

3 PM

4 PM

5 PM

6 PM

Delays / Problems

Schedule Updates / Progress

Extra Work / Authorized by

Supervisor's Signature ______________________

JOB NAME: ______________________ DATE: ______________

CONTRACTOR: ______________________ Weather: AM ______ PM ______

Expenses / Materials

____

____

____

____

____

Material Deliveries

____

____

____

____

____

Equipment Use / Hours

____

____

____

____

____

Equipment Rentals

____

____

____

____

____

Daily Work Force	**No.**
Superintendent	______
Bricklayers	______
Carpenters	______
Masons	______
Electricians	______
Iron Workers	______
Plumbers	______
Others	______
______________	______
Total	______

DAILY WORK LOG

7 AM ______________________________

____

8 AM ______________________________

____

9 AM ______________________________

____

10 AM ______________________________

____

11 AM ______________________________

____

12 NOON ______________________________

____

1 PM ______________________________

____

2 PM ______________________________

____

3 PM ______________________________

____

4 PM ______________________________

____

5 PM ______________________________

____

6 PM ______________________________

____

Delays / Problems ______________________________

____

____

Schedule Updates / Progress ______________________________

____

____

Extra Work / Authorized by ______________________________

____

Supervisor's Signature ______________________________

JOB NAME: ____________________ DATE: __________

CONTRACTOR: ____________________ Weather: AM ______ PM ______

DAILY WORK LOG

Expenses / Materials

Material Deliveries

Equipment Use / Hours

Equipment Rentals

Daily Work Force **No.**

Superintendent	______
Bricklayers	______
Carpenters	______
Masons	______
Electricians	______
Iron Workers	______
Plumbers	______
Others	______
______	______
Total	______

7 AM

8 AM

9 AM

10 AM

11 AM

12 NOON

1 PM

2 PM

3 PM

4 PM

5 PM

6 PM

Delays / Problems

Schedule Updates / Progress

Extra Work / Authorized by

Supervisor's Signature ____________________

JOB NAME: ______________________________ DATE: ______________

CONTRACTOR: ______________________________ Weather: AM ______ PM ______

Expenses / Materials	**DAILY WORK LOG**
______________	7 AM ______________
______________	______________
______________	8 AM ______________
______________	______________
______________	9 AM ______________
Material Deliveries	______________
______________	10 AM ______________
______________	______________
______________	11 AM ______________
______________	______________
______________	12 NOON ______________
Equipment Use / Hours	______________
______________	1 PM ______________
______________	______________
______________	2 PM ______________
______________	______________
______________	3 PM ______________
Equipment Rentals	______________
______________	4 PM ______________
______________	______________
______________	5 PM ______________
______________	______________
______________	6 PM ______________
Daily Work Force **No.**	______________
Superintendent ______	Delays / Problems ______________
Bricklayers ______	______________
Carpenters ______	______________
Masons ______	Schedule Updates / Progress ______________
Electricians ______	______________
Iron Workers ______	______________
Plumbers ______	Extra Work / Authorized by ______________
Others ______	______________
______________ ______	
Total ______	

Supervisor's Signature ______________________________

JOB NAME: ____________________ DATE: ____________

CONTRACTOR: ____________________ Weather: AM ______ PM ______

DAILY WORK LOG

Expenses / Materials

Material Deliveries

Equipment Use / Hours

Equipment Rentals

Daily Work Force **No.**

Superintendent ______

Bricklayers ______

Carpenters ______

Masons ______

Electricians ______

Iron Workers ______

Plumbers ______

Others ______

____________ ______

Total ______

7 AM

8 AM

9 AM

10 AM

11 AM

12 NOON

1 PM

2 PM

3 PM

4 PM

5 PM

6 PM

Delays / Problems

Schedule Updates / Progress

Extra Work / Authorized by

Supervisor's Signature ____________________

JOB NAME: ______________________ DATE: ______________

CONTRACTOR: ______________________ Weather: AM ______ PM ______

DAILY WORK LOG

Expenses / Materials

____

____

____

____

____

Material Deliveries

____

____

____

____

____

Equipment Use / Hours

____

____

____

____

____

Equipment Rentals

____

____

____

____

____

Daily Work Force	**No.**
Superintendent	______
Bricklayers	______
Carpenters	______
Masons	______
Electricians	______
Iron Workers	______
Plumbers	______
Others	______
______________	______
Total	______

7 AM ______________________________

____

8 AM ______________________________

____

9 AM ______________________________

____

10 AM ______________________________

____

11 AM ______________________________

____

12 NOON ______________________________

____

1 PM ______________________________

____

2 PM ______________________________

____

3 PM ______________________________

____

4 PM ______________________________

____

5 PM ______________________________

____

6 PM ______________________________

____

Delays / Problems ______________________________

____

____

Schedule Updates / Progress ______________________________

____

____

Extra Work / Authorized by ______________________________

____

Supervisor's Signature ______________________________

JOB NAME: ______________________________ DATE: ______________

CONTRACTOR: ______________________________ Weather: AM ________ PM ________

DAILY WORK LOG

Expenses / Materials

____

____

____

____

____

Material Deliveries

____

____

____

____

____

Equipment Use / Hours

____

____

____

____

____

Equipment Rentals

____

____

____

____

____

Daily Work Force **No.**

Superintendent	______
Bricklayers	______
Carpenters	______
Masons	______
Electricians	______
Iron Workers	______
Plumbers	______
Others	______
____________	______
Total	______

7 AM ______________________________

____

8 AM ______________________________

____

9 AM ______________________________

____

10 AM ______________________________

____

11 AM ______________________________

____

12 NOON ______________________________

____

1 PM ______________________________

____

2 PM ______________________________

____

3 PM ______________________________

____

4 PM ______________________________

____

5 PM ______________________________

____

6 PM ______________________________

____

Delays / Problems ______________________________

____

____

Schedule Updates / Progress ______________________________

____

____

Extra Work / Authorized by ______________________________

____

Supervisor's Signature ______________________________

JOB NAME: ______________________ DATE: ____________

CONTRACTOR: ______________________ Weather: AM ________ PM ________

Expenses / Materials

Material Deliveries

Equipment Use / Hours

Equipment Rentals

Daily Work Force **No.**

Superintendent	____
Bricklayers	____
Carpenters	____
Masons	____
Electricians	____
Iron Workers	____
Plumbers	____
Others	____
____________	____
Total	____

DAILY WORK LOG

7 AM

8 AM

9 AM

10 AM

11 AM

12 NOON

1 PM

2 PM

3 PM

4 PM

5 PM

6 PM

Delays / Problems

Schedule Updates / Progress

Extra Work / Authorized by

Supervisor's Signature ______________________

JOB NAME: ______________________ DATE: ______________

CONTRACTOR: ______________________ Weather: AM ______ PM ______

DAILY WORK LOG

Expenses / Materials

____

Material Deliveries

____

Equipment Use / Hours

____

Equipment Rentals

____

Daily Work Force **No.**

Superintendent ______

Bricklayers ______

Carpenters ______

Masons ______

Electricians ______

Iron Workers ______

Plumbers ______

Others ______

______________ ______

Total ______

7 AM ______________________

8 AM ______________________

9 AM ______________________

10 AM ______________________

11 AM ______________________

12 NOON ______________________

1 PM ______________________

2 PM ______________________

3 PM ______________________

4 PM ______________________

5 PM ______________________

6 PM ______________________

Delays / Problems ______________________

Schedule Updates / Progress ______________________

Extra Work / Authorized by ______________________

Supervisor's Signature ______________________

JOB NAME: ____________________ DATE: ____________

CONTRACTOR: ____________________ Weather: AM ______ PM ______

DAILY WORK LOG

Expenses / Materials

Material Deliveries

Equipment Use / Hours

Equipment Rentals

Daily Work Force **No.**

Daily Work Force	No.
Superintendent	____
Bricklayers	____
Carpenters	____
Masons	____
Electricians	____
Iron Workers	____
Plumbers	____
Others	____
________	____
Total	____

7 AM

8 AM

9 AM

10 AM

11 AM

12 NOON

1 PM

2 PM

3 PM

4 PM

5 PM

6 PM

Delays / Problems

Schedule Updates / Progress

Extra Work / Authorized by

Supervisor's Signature ____________________

JOB NAME: ______________________ DATE: ____________

CONTRACTOR: ______________________ Weather: AM ________ PM ________

DAILY WORK LOG

Expenses / Materials

Material Deliveries

Equipment Use / Hours

Equipment Rentals

Daily Work Force	**No.**
Superintendent	
Bricklayers	
Carpenters	
Masons	
Electricians	
Iron Workers	
Plumbers	
Others	
Total	

7 AM

8 AM

9 AM

10 AM

11 AM

12 NOON

1 PM

2 PM

3 PM

4 PM

5 PM

6 PM

Delays / Problems

Schedule Updates / Progress

Extra Work / Authorized by

Supervisor's Signature ______________________

JOB NAME: ______________________ DATE: ______________

CONTRACTOR: ______________________ Weather: AM ________ PM ________

Expenses / Materials

Material Deliveries

Equipment Use / Hours

Equipment Rentals

Daily Work Force **No.**

Superintendent	
Bricklayers	
Carpenters	
Masons	
Electricians	
Iron Workers	
Plumbers	
Others	
Total	

DAILY WORK LOG

7 AM

8 AM

9 AM

10 AM

11 AM

12 NOON

1 PM

2 PM

3 PM

4 PM

5 PM

6 PM

Delays / Problems

Schedule Updates / Progress

Extra Work / Authorized by

Supervisor's Signature ______________________

JOB NAME: ______________________ DATE: ______________

CONTRACTOR: ______________________ Weather: AM ________ PM ________

DAILY WORK LOG

Expenses / Materials

____

____

____

____

____

Material Deliveries

____

____

____

____

____

Equipment Use / Hours

____

____

____

____

____

Equipment Rentals

____

____

____

____

____

Daily Work Force	**No.**
Superintendent	______
Bricklayers	______
Carpenters	______
Masons	______
Electricians	______
Iron Workers	______
Plumbers	______
Others	______
______________	______
Total	______

7 AM ______________________________

8 AM ______________________________

9 AM ______________________________

10 AM ______________________________

11 AM ______________________________

12 NOON ______________________________

1 PM ______________________________

2 PM ______________________________

3 PM ______________________________

4 PM ______________________________

5 PM ______________________________

6 PM ______________________________

Delays / Problems ______________________________

Schedule Updates / Progress ______________________________

Extra Work / Authorized by ______________________________

Supervisor's Signature ______________________________

JOB NAME: ______________________ DATE: ______________

CONTRACTOR: ______________________ Weather: AM ______ PM ______

Expenses / Materials

Material Deliveries

Equipment Use / Hours

Equipment Rentals

Daily Work Force No.

Superintendent ______

Bricklayers ______

Carpenters ______

Masons ______

Electricians ______

Iron Workers ______

Plumbers ______

Others ______

______ ______

Total ______

DAILY WORK LOG

7 AM

8 AM

9 AM

10 AM

11 AM

12 NOON

1 PM

2 PM

3 PM

4 PM

5 PM

6 PM

Delays / Problems

Schedule Updates / Progress

Extra Work / Authorized by

Supervisor's Signature ______________________

JOB NAME: ______________________ DATE: ____________

CONTRACTOR: ______________________ Weather: AM ______ PM ______

DAILY WORK LOG

Expenses / Materials

Material Deliveries

Equipment Use / Hours

Equipment Rentals

Daily Work Force	**No.**
Superintendent	
Bricklayers	
Carpenters	
Masons	
Electricians	
Iron Workers	
Plumbers	
Others	
Total	

7 AM

8 AM

9 AM

10 AM

11 AM

12 NOON

1 PM

2 PM

3 PM

4 PM

5 PM

6 PM

Delays / Problems

Schedule Updates / Progress

Extra Work / Authorized by

Supervisor's Signature ______________________

JOB NAME: ______________________ DATE: ______________

CONTRACTOR: ______________________ Weather: AM ______ PM ______

Expenses / Materials

____

Material Deliveries

____

Equipment Use / Hours

____

Equipment Rentals

____

Daily Work Force No.

Superintendent	______
Bricklayers	______
Carpenters	______
Masons	______
Electricians	______
Iron Workers	______
Plumbers	______
Others	______
______________	______
Total	______

DAILY WORK LOG

7 AM ______________________

8 AM ______________________

9 AM ______________________

10 AM ______________________

11 AM ______________________

12 NOON ______________________

1 PM ______________________

2 PM ______________________

3 PM ______________________

4 PM ______________________

5 PM ______________________

6 PM ______________________

Delays / Problems ______________________

Schedule Updates / Progress ______________________

Extra Work / Authorized by ______________________

Supervisor's Signature ______________________

JOB NAME: ______________________ DATE: ______________

CONTRACTOR: ______________________ Weather: AM ______ PM ______

DAILY WORK LOG

Expenses / Materials

____

Material Deliveries

____

Equipment Use / Hours

____

Equipment Rentals

____

Daily Work Force	**No.**
Superintendent	______
Bricklayers	______
Carpenters	______
Masons	______
Electricians	______
Iron Workers	______
Plumbers	______
Others	______
____________	______
Total	______

7 AM ______________________

8 AM ______________________

9 AM ______________________

10 AM ______________________

11 AM ______________________

12 NOON ______________________

1 PM ______________________

2 PM ______________________

3 PM ______________________

4 PM ______________________

5 PM ______________________

6 PM ______________________

Delays / Problems ______________________

Schedule Updates / Progress ______________________

Extra Work / Authorized by ______________________

Supervisor's Signature ______________________

JOB NAME: ______________________ DATE: ______________

CONTRACTOR: ______________________ Weather: AM ______ PM ______

DAILY WORK LOG

Expenses / Materials

Material Deliveries

Equipment Use / Hours

Equipment Rentals

Daily Work Force	**No.**
Superintendent	
Bricklayers	
Carpenters	
Masons	
Electricians	
Iron Workers	
Plumbers	
Others	
Total	

7 AM

8 AM

9 AM

10 AM

11 AM

12 NOON

1 PM

2 PM

3 PM

4 PM

5 PM

6 PM

Delays / Problems

Schedule Updates / Progress

Extra Work / Authorized by

Supervisor's Signature ______________________

JOB NAME: ______________________ DATE: __________

CONTRACTOR: ______________________ Weather: AM ______ PM ______

DAILY WORK LOG

Expenses / Materials

Material Deliveries

Equipment Use / Hours

Equipment Rentals

Daily Work Force **No.**

Superintendent ______

Bricklayers ______

Carpenters ______

Masons ______

Electricians ______

Iron Workers ______

Plumbers ______

Others ______

__________ ______

Total ______

7 AM

8 AM

9 AM

10 AM

11 AM

12 NOON

1 PM

2 PM

3 PM

4 PM

5 PM

6 PM

Delays / Problems

Schedule Updates / Progress

Extra Work / Authorized by

Supervisor's Signature ______________________

JOB NAME: ______________________ DATE: ______________

CONTRACTOR: ______________________ Weather: AM ______ PM ______

Expenses / Materials

____

____

____

____

____

Material Deliveries

____

____

____

____

____

Equipment Use / Hours

____

____

____

____

____

Equipment Rentals

____

____

____

____

____

Daily Work Force	**No.**
Superintendent	______
Bricklayers	______
Carpenters	______
Masons	______
Electricians	______
Iron Workers	______
Plumbers	______
Others	______
____________	______
Total	______

DAILY WORK LOG

7 AM ______________________________

____

8 AM ______________________________

____

9 AM ______________________________

____

10 AM ______________________________

____

11 AM ______________________________

____

12 NOON ______________________________

____

1 PM ______________________________

____

2 PM ______________________________

____

3 PM ______________________________

____

4 PM ______________________________

____

5 PM ______________________________

____

6 PM ______________________________

____

Delays / Problems ______________________________

____

____

Schedule Updates / Progress ______________________________

____

____

Extra Work / Authorized by ______________________________

____

Supervisor's Signature ______________________________

JOB NAME: ______________________ DATE: ____________

CONTRACTOR: ______________________ Weather: AM ________ PM ________

DAILY WORK LOG

Expenses / Materials

Material Deliveries

Equipment Use / Hours

Equipment Rentals

Daily Work Force **No.**

Superintendent ____

Bricklayers ____

Carpenters ____

Masons ____

Electricians ____

Iron Workers ____

Plumbers ____

Others ____

________ ____

Total ____

7 AM

8 AM

9 AM

10 AM

11 AM

12 NOON

1 PM

2 PM

3 PM

4 PM

5 PM

6 PM

Delays / Problems

Schedule Updates / Progress

Extra Work / Authorized by

Supervisor's Signature ______________________

JOB NAME: ______________________ DATE: ______________

CONTRACTOR: ______________________ Weather: AM ______ PM ______

DAILY WORK LOG

Expenses / Materials

____

Material Deliveries

____

Equipment Use / Hours

____

Equipment Rentals

____

Daily Work Force	**No.**
Superintendent	______
Bricklayers	______
Carpenters	______
Masons	______
Electricians	______
Iron Workers	______
Plumbers	______
Others	______
______________	______
Total	______

7 AM ______________________

8 AM ______________________

9 AM ______________________

10 AM ______________________

11 AM ______________________

12 NOON ______________________

1 PM ______________________

2 PM ______________________

3 PM ______________________

4 PM ______________________

5 PM ______________________

6 PM ______________________

Delays / Problems ______________________

Schedule Updates / Progress ______________________

Extra Work / Authorized by ______________________

Supervisor's Signature ______________________

JOB NAME: ______________________ DATE: ______________
CONTRACTOR: ______________________ Weather: AM ______ PM ______

DAILY WORK LOG

Expenses / Materials

Material Deliveries

Equipment Use / Hours

Equipment Rentals

Daily Work Force **No.**

Superintendent	
Bricklayers	
Carpenters	
Masons	
Electricians	
Iron Workers	
Plumbers	
Others	
Total	

7 AM

8 AM

9 AM

10 AM

11 AM

12 NOON

1 PM

2 PM

3 PM

4 PM

5 PM

6 PM

Delays / Problems

Schedule Updates / Progress

Extra Work / Authorized by

Supervisor's Signature ______________________

JOB NAME: ______________________ DATE: ____________

CONTRACTOR: ______________________ Weather: AM ______ PM ______

Expenses / Materials

____

____

____

____

____

Material Deliveries

____

____

____

____

____

Equipment Use / Hours

____

____

____

____

____

Equipment Rentals

____

____

____

____

____

Daily Work Force	**No.**
Superintendent	______
Bricklayers	______
Carpenters	______
Masons	______
Electricians	______
Iron Workers	______
Plumbers	______
Others	______
____________	______
Total	______

DAILY WORK LOG

7 AM ______________________________

____

8 AM ______________________________

____

9 AM ______________________________

____

10 AM ______________________________

____

11 AM ______________________________

____

12 NOON ______________________________

____

1 PM ______________________________

____

2 PM ______________________________

____

3 PM ______________________________

____

4 PM ______________________________

____

5 PM ______________________________

____

6 PM ______________________________

____

Delays / Problems ______________________________

____

____

Schedule Updates / Progress ______________________________

____

____

Extra Work / Authorized by ______________________________

____

Supervisor's Signature ______________________________

JOB NAME: ______________________ DATE: ______________

CONTRACTOR: ______________________ Weather: AM ________ PM ________

DAILY WORK LOG

Expenses / Materials

Material Deliveries

Equipment Use / Hours

Equipment Rentals

Daily Work Force **No.**

Superintendent ______

Bricklayers ______

Carpenters ______

Masons ______

Electricians ______

Iron Workers ______

Plumbers ______

Others ______

____________ ______

Total ______

7 AM

8 AM

9 AM

10 AM

11 AM

12 NOON

1 PM

2 PM

3 PM

4 PM

5 PM

6 PM

Delays / Problems

Schedule Updates / Progress

Extra Work / Authorized by

Supervisor's Signature ______________________________

JOB NAME: ______________________ DATE: ______________

CONTRACTOR: ______________________ Weather: AM ________ PM ________

DAILY WORK LOG

Expenses / Materials	

Material Deliveries

Equipment Use / Hours

Equipment Rentals

Daily Work Force **No.**

Superintendent ______

Bricklayers ______

Carpenters ______

Masons ______

Electricians ______

Iron Workers ______

Plumbers ______

Others ______

____________ ______

Total ______

7 AM

8 AM

9 AM

10 AM

11 AM

12 NOON

1 PM

2 PM

3 PM

4 PM

5 PM

6 PM

Delays / Problems

Schedule Updates / Progress

Extra Work / Authorized by

Supervisor's Signature ______________________________

JOB NAME: ______________________ DATE: ____________

CONTRACTOR: ______________________ Weather: AM ______ PM ______

DAILY WORK LOG

Expenses / Materials

Material Deliveries

Equipment Use / Hours

Equipment Rentals

Daily Work Force **No.**

Superintendent	____
Bricklayers	____
Carpenters	____
Masons	____
Electricians	____
Iron Workers	____
Plumbers	____
Others	____
________	____
Total	____

7 AM

8 AM

9 AM

10 AM

11 AM

12 NOON

1 PM

2 PM

3 PM

4 PM

5 PM

6 PM

Delays / Problems

Schedule Updates / Progress

Extra Work / Authorized by

Supervisor's Signature ______________________

JOB NAME: ______________________ DATE: ______________

CONTRACTOR: ______________________ Weather: AM ______ PM ______

DAILY WORK LOG

Expenses / Materials	
	7 AM
	8 AM
Material Deliveries	9 AM
	10 AM
	11 AM
Equipment Use / Hours	12 NOON
	1 PM
	2 PM
Equipment Rentals	3 PM
	4 PM
	5 PM
Daily Work Force — No.	6 PM
Superintendent ____	
Bricklayers ____	
Carpenters ____	Delays / Problems
Masons ____	
Electricians ____	
Iron Workers ____	Schedule Updates / Progress
Plumbers ____	
Others ____	
________ ____	Extra Work / Authorized by
Total ____	

Supervisor's Signature ______________________________

JOB NAME: ______________________ DATE: ______________

CONTRACTOR: ______________________ Weather: AM ________ PM ________

Expenses / Materials

Material Deliveries

Equipment Use / Hours

Equipment Rentals

Daily Work Force No.

Superintendent ______

Bricklayers ______

Carpenters ______

Masons ______

Electricians ______

Iron Workers ______

Plumbers ______

Others ______

______ ______

Total ______

DAILY WORK LOG

7 AM

8 AM

9 AM

10 AM

11 AM

12 NOON

1 PM

2 PM

3 PM

4 PM

5 PM

6 PM

Delays / Problems

Schedule Updates / Progress

Extra Work / Authorized by

Supervisor's Signature ______________________

JOB NAME: ______________________ DATE: ______________

CONTRACTOR: ______________________ Weather: AM ______ PM ______

Expenses / Materials

____
____
____
____
____

Material Deliveries

____
____
____
____
____

Equipment Use / Hours

____
____
____
____
____

Equipment Rentals

____
____
____
____
____

Daily Work Force	**No.**
Superintendent	____
Bricklayers	____
Carpenters	____
Masons	____
Electricians	____
Iron Workers	____
Plumbers	____
Others	____
__________	____
Total	____

DAILY WORK LOG

7 AM ______________________

8 AM ______________________

9 AM ______________________

10 AM ______________________

11 AM ______________________

12 NOON ______________________

1 PM ______________________

2 PM ______________________

3 PM ______________________

4 PM ______________________

5 PM ______________________

6 PM ______________________

Delays / Problems ______________________

Schedule Updates / Progress ______________________

Extra Work / Authorized by ______________________

Supervisor's Signature ______________________

JOB NAME: ______________________ DATE: __________

CONTRACTOR: ______________________ Weather: AM ______ PM ______

Expenses / Materials

Material Deliveries

Equipment Use / Hours

Equipment Rentals

Daily Work Force **No.**

Superintendent	____
Bricklayers	____
Carpenters	____
Masons	____
Electricians	____
Iron Workers	____
Plumbers	____
Others	____
________	____
Total	____

DAILY WORK LOG

7 AM

8 AM

9 AM

10 AM

11 AM

12 NOON

1 PM

2 PM

3 PM

4 PM

5 PM

6 PM

Delays / Problems

Schedule Updates / Progress

Extra Work / Authorized by

Supervisor's Signature ______________________

JOB NAME: ______________________ DATE: ______________

CONTRACTOR: ______________________ Weather: AM ______ PM ______

Expenses / Materials

Material Deliveries

Equipment Use / Hours

Equipment Rentals

Daily Work Force **No.**

Daily Work Force	No.
Superintendent	
Bricklayers	
Carpenters	
Masons	
Electricians	
Iron Workers	
Plumbers	
Others	
Total	

DAILY WORK LOG

7 AM

8 AM

9 AM

10 AM

11 AM

12 NOON

1 PM

2 PM

3 PM

4 PM

5 PM

6 PM

Delays / Problems

Schedule Updates / Progress

Extra Work / Authorized by

Supervisor's Signature ______________________

JOB NAME: ______________________ DATE: ____________

CONTRACTOR: ______________________ Weather: AM ______ PM ______

DAILY WORK LOG

Expenses / Materials

Material Deliveries

Equipment Use / Hours

Equipment Rentals

Daily Work Force **No.**

Superintendent ______
Bricklayers ______
Carpenters ______
Masons ______
Electricians ______
Iron Workers ______
Plumbers ______
Others ______
______ ______

Total ______

7 AM

8 AM

9 AM

10 AM

11 AM

12 NOON

1 PM

2 PM

3 PM

4 PM

5 PM

6 PM

Delays / Problems

Schedule Updates / Progress

Extra Work / Authorized by

Supervisor's Signature ______________________

JOB NAME: ______________________ DATE: ______________

CONTRACTOR: ______________________ Weather: AM _______ PM _______

DAILY WORK LOG

Expenses / Materials

Material Deliveries

Equipment Use / Hours

Equipment Rentals

Daily Work Force **No.**

Daily Work Force	No.
Superintendent	
Bricklayers	
Carpenters	
Masons	
Electricians	
Iron Workers	
Plumbers	
Others	
Total	

7 AM

8 AM

9 AM

10 AM

11 AM

12 NOON

1 PM

2 PM

3 PM

4 PM

5 PM

6 PM

Delays / Problems

Schedule Updates / Progress

Extra Work / Authorized by

Supervisor's Signature ______________________

JOB NAME: ____________________ DATE: __________

CONTRACTOR: ____________________ Weather: AM ______ PM ______

DAILY WORK LOG

Expenses / Materials

Material Deliveries

Equipment Use / Hours

Equipment Rentals

Daily Work Force **No.**

Superintendent	____
Bricklayers	____
Carpenters	____
Masons	____
Electricians	____
Iron Workers	____
Plumbers	____
Others	____
______	____
Total	____

7 AM

8 AM

9 AM

10 AM

11 AM

12 NOON

1 PM

2 PM

3 PM

4 PM

5 PM

6 PM

Delays / Problems

Schedule Updates / Progress

Extra Work / Authorized by

Supervisor's Signature ____________________

JOB NAME: ______________________ DATE: ______________

CONTRACTOR: ______________________ Weather: AM ________ PM ________

DAILY WORK LOG

Expenses / Materials	
	7 AM
	8 AM
Material Deliveries	9 AM
	10 AM
	11 AM
Equipment Use / Hours	12 NOON
	1 PM
	2 PM
Equipment Rentals	3 PM
	4 PM
	5 PM
Daily Work Force **No.**	6 PM
Superintendent ____	
Bricklayers ____	
Carpenters ____	Delays / Problems
Masons ____	
Electricians ____	
Iron Workers ____	Schedule Updates / Progress
Plumbers ____	
Others ____	
________ ____	Extra Work / Authorized by
Total ____	

Supervisor's Signature ______________________________

JOB NAME: ______________________ DATE: ____________

CONTRACTOR: ______________________ Weather: AM ________ PM ________

DAILY WORK LOG

Expenses / Materials

Material Deliveries

Equipment Use / Hours

Equipment Rentals

Daily Work Force **No.**

Superintendent	____
Bricklayers	____
Carpenters	____
Masons	____
Electricians	____
Iron Workers	____
Plumbers	____
Others	____
________	____
Total	____

7 AM

8 AM

9 AM

10 AM

11 AM

12 NOON

1 PM

2 PM

3 PM

4 PM

5 PM

6 PM

Delays / Problems

Schedule Updates / Progress

Extra Work / Authorized by

Supervisor's Signature ______________________

JOB NAME: ______________________ DATE: ______________

CONTRACTOR: ______________________ Weather: AM ______ PM ______

DAILY WORK LOG

Expenses / Materials

____

____

____

____

____

Material Deliveries

____

____

____

____

____

Equipment Use / Hours

____

____

____

____

____

Equipment Rentals

____

____

____

____

____

Daily Work Force	**No.**
Superintendent	______
Bricklayers	______
Carpenters	______
Masons	______
Electricians	______
Iron Workers	______
Plumbers	______
Others	______
______________	______
Total	______

7 AM ______________________________

____

8 AM ______________________________

____

9 AM ______________________________

____

10 AM ______________________________

____

11 AM ______________________________

____

12 NOON ______________________________

____

1 PM ______________________________

____

2 PM ______________________________

____

3 PM ______________________________

____

4 PM ______________________________

____

5 PM ______________________________

____

6 PM ______________________________

____

Delays / Problems ______________________________

____

____

Schedule Updates / Progress ______________________________

____

____

Extra Work / Authorized by ______________________________

____

Supervisor's Signature ______________________________

JOB NAME: ______________________ DATE: ______________

CONTRACTOR: ______________________ Weather: AM ______ PM ______

Expenses / Materials

Material Deliveries

Equipment Use / Hours

Equipment Rentals

Daily Work Force **No.**

Superintendent ______

Bricklayers ______

Carpenters ______

Masons ______

Electricians ______

Iron Workers ______

Plumbers ______

Others ______

______________ ______

Total ______

DAILY WORK LOG

7 AM

8 AM

9 AM

10 AM

11 AM

12 NOON

1 PM

2 PM

3 PM

4 PM

5 PM

6 PM

Delays / Problems

Schedule Updates / Progress

Extra Work / Authorized by

Supervisor's Signature ______________________________

JOB NAME: ______________________ DATE: ____________

CONTRACTOR: ______________________ Weather: AM ______ PM ______

Expenses / Materials

Material Deliveries

Equipment Use / Hours

Equipment Rentals

Daily Work Force	**No.**
Superintendent	
Bricklayers	
Carpenters	
Masons	
Electricians	
Iron Workers	
Plumbers	
Others	
Total	

DAILY WORK LOG

7 AM

8 AM

9 AM

10 AM

11 AM

12 NOON

1 PM

2 PM

3 PM

4 PM

5 PM

6 PM

Delays / Problems

Schedule Updates / Progress

Extra Work / Authorized by

Supervisor's Signature ______________________

JOB NAME: ______________________ DATE: ______________

CONTRACTOR: ______________________ Weather: AM ________ PM ________

DAILY WORK LOG

Expenses / Materials

Material Deliveries

Equipment Use / Hours

Equipment Rentals

Daily Work Force **No.**

Superintendent	____
Bricklayers	____
Carpenters	____
Masons	____
Electricians	____
Iron Workers	____
Plumbers	____
Others	____
________	____
Total	____

7 AM

8 AM

9 AM

10 AM

11 AM

12 NOON

1 PM

2 PM

3 PM

4 PM

5 PM

6 PM

Delays / Problems

Schedule Updates / Progress

Extra Work / Authorized by

Supervisor's Signature ______________________

JOB NAME: ______________________ DATE: ______________

CONTRACTOR: ______________________ Weather: AM ______ PM ______

DAILY WORK LOG

Expenses / Materials	
	7 AM
	8 AM
Material Deliveries	9 AM
	10 AM
	11 AM
Equipment Use / Hours	12 NOON
	1 PM
	2 PM
Equipment Rentals	3 PM
	4 PM
	5 PM

Daily Work Force **No.**

Superintendent	______
Bricklayers	______
Carpenters	______
Masons	______
Electricians	______
Iron Workers	______
Plumbers	______
Others	______
____________	______
Total	______

6 PM

Delays / Problems

Schedule Updates / Progress

Extra Work / Authorized by

Supervisor's Signature ______________________________

JOB NAME: ______________________ DATE: ____________

CONTRACTOR: ______________________ Weather: AM ________ PM ________

DAILY WORK LOG

Expenses / Materials

Material Deliveries

Equipment Use / Hours

Equipment Rentals

Daily Work Force	**No.**
Superintendent	
Bricklayers	
Carpenters	
Masons	
Electricians	
Iron Workers	
Plumbers	
Others	
Total	

7 AM

8 AM

9 AM

10 AM

11 AM

12 NOON

1 PM

2 PM

3 PM

4 PM

5 PM

6 PM

Delays / Problems

Schedule Updates / Progress

Extra Work / Authorized by

Supervisor's Signature ______________________

JOB NAME: ______________________ DATE: ____________

CONTRACTOR: ______________________ Weather: AM ______ PM ______

DAILY WORK LOG

Expenses / Materials

Material Deliveries

Equipment Use / Hours

Equipment Rentals

Daily Work Force **No.**

Superintendent ______
Bricklayers ______
Carpenters ______
Masons ______
Electricians ______
Iron Workers ______
Plumbers ______
Others ______
______ ______

Total ______

7 AM

8 AM

9 AM

10 AM

11 AM

12 NOON

1 PM

2 PM

3 PM

4 PM

5 PM

6 PM

Delays / Problems

Schedule Updates / Progress

Extra Work / Authorized by

Supervisor's Signature ______________________

JOB NAME: ______________________ DATE: __________

CONTRACTOR: ______________________ Weather: AM ______ PM ______

DAILY WORK LOG

Expenses / Materials

Material Deliveries

Equipment Use / Hours

Equipment Rentals

Daily Work Force	**No.**
Superintendent	____
Bricklayers	____
Carpenters	____
Masons	____
Electricians	____
Iron Workers	____
Plumbers	____
Others	____
____________	____
Total	____

7 AM

8 AM

9 AM

10 AM

11 AM

12 NOON

1 PM

2 PM

3 PM

4 PM

5 PM

6 PM

Delays / Problems

Schedule Updates / Progress

Extra Work / Authorized by

Supervisor's Signature ______________________

JOB NAME: ______________________________ DATE: ______________

CONTRACTOR: ______________________________ Weather: AM ______ PM ______

Expenses / Materials

____
____
____
____
____

Material Deliveries

____
____
____
____
____

Equipment Use / Hours

____
____
____
____
____

Equipment Rentals

____
____
____
____
____

Daily Work Force	**No.**
Superintendent	______
Bricklayers	______
Carpenters	______
Masons	______
Electricians	______
Iron Workers	______
Plumbers	______
Others	______
____________	______
Total	______

DAILY WORK LOG

7 AM ______________________________

____

8 AM ______________________________

____

9 AM ______________________________

____

10 AM ______________________________

____

11 AM ______________________________

____

12 NOON ______________________________

____

1 PM ______________________________

____

2 PM ______________________________

____

3 PM ______________________________

____

4 PM ______________________________

____

5 PM ______________________________

____

6 PM ______________________________

____

Delays / Problems ______________________________

____

____

Schedule Updates / Progress ______________________________

____

____

Extra Work / Authorized by ______________________________

____

Supervisor's Signature ______________________________

JOB NAME: ______________________ DATE: ______________

CONTRACTOR: ______________________ Weather: AM ________ PM ________

Expenses / Materials

DAILY WORK LOG

7 AM

8 AM

9 AM

10 AM

11 AM

12 NOON

1 PM

2 PM

3 PM

4 PM

5 PM

6 PM

Delays / Problems

Schedule Updates / Progress

Extra Work / Authorized by

Material Deliveries

Equipment Use / Hours

Equipment Rentals

Daily Work Force	**No.**
Superintendent	____
Bricklayers	____
Carpenters	____
Masons	____
Electricians	____
Iron Workers	____
Plumbers	____
Others	____
____________	____
Total	____

Supervisor's Signature ______________________

JOB NAME: ______________________ DATE: ____________

CONTRACTOR: ______________________ Weather: AM ______ PM ______

Expenses / Materials	**DAILY WORK LOG**
	7 AM
	8 AM
Material Deliveries	9 AM
	10 AM
	11 AM
Equipment Use / Hours	12 NOON
	1 PM
	2 PM
Equipment Rentals	3 PM
	4 PM
	5 PM
Daily Work Force **No.**	6 PM
Superintendent ______	
Bricklayers ______	Delays / Problems
Carpenters ______	
Masons ______	
Electricians ______	Schedule Updates / Progress
Iron Workers ______	
Plumbers ______	
Others ______	Extra Work / Authorized by
____________ ______	
Total ______	

Supervisor's Signature ______________________

JOB NAME: ______________________ DATE: ______________

CONTRACTOR: ______________________ Weather: AM ________ PM ________

DAILY WORK LOG

Expenses / Materials

Material Deliveries

Equipment Use / Hours

Equipment Rentals

Daily Work Force **No.**

Superintendent	______
Bricklayers	______
Carpenters	______
Masons	______
Electricians	______
Iron Workers	______
Plumbers	______
Others	______
______________	______
Total	______

7 AM

8 AM

9 AM

10 AM

11 AM

12 NOON

1 PM

2 PM

3 PM

4 PM

5 PM

6 PM

Delays / Problems

Schedule Updates / Progress

Extra Work / Authorized by

Supervisor's Signature ______________________

JOB NAME: ______________________ DATE: ______________

CONTRACTOR: ______________________ Weather: AM ________ PM ________

Expenses / Materials

____

____

____

____

____

Material Deliveries

____

____

____

____

____

Equipment Use / Hours

____

____

____

____

____

Equipment Rentals

____

____

____

____

____

Daily Work Force	**No.**
Superintendent	______
Bricklayers	______
Carpenters	______
Masons	______
Electricians	______
Iron Workers	______
Plumbers	______
Others	______
______________	______
Total	______

DAILY WORK LOG

7 AM ______________________________

____

8 AM ______________________________

____

9 AM ______________________________

____

10 AM ______________________________

____

11 AM ______________________________

____

12 NOON ______________________________

____

1 PM ______________________________

____

2 PM ______________________________

____

3 PM ______________________________

____

4 PM ______________________________

____

5 PM ______________________________

____

6 PM ______________________________

____

Delays / Problems ______________________________

____

____

Schedule Updates / Progress ______________________________

____

____

Extra Work / Authorized by ______________________________

____

Supervisor's Signature ______________________________

JOB NAME: ______________________ DATE: ______________

CONTRACTOR: ______________________ Weather: AM ______ PM ______

DAILY WORK LOG

Expenses / Materials

Material Deliveries

Equipment Use / Hours

Equipment Rentals

Daily Work Force	No.
Superintendent	
Bricklayers	
Carpenters	
Masons	
Electricians	
Iron Workers	
Plumbers	
Others	
Total	

7 AM

8 AM

9 AM

10 AM

11 AM

12 NOON

1 PM

2 PM

3 PM

4 PM

5 PM

6 PM

Delays / Problems

Schedule Updates / Progress

Extra Work / Authorized by

Supervisor's Signature ______________________

JOB NAME: ______________________ DATE: ____________

CONTRACTOR: ______________________ Weather: AM ______ PM ______

DAILY WORK LOG

Expenses / Materials

____

____

____

____

____

Material Deliveries

____

____

____

____

____

Equipment Use / Hours

____

____

____

____

____

Equipment Rentals

____

____

____

____

____

Daily Work Force	**No.**
Superintendent	______
Bricklayers	______
Carpenters	______
Masons	______
Electricians	______
Iron Workers	______
Plumbers	______
Others	______
______________	______
Total	______

7 AM ______________________

____

8 AM ______________________

____

9 AM ______________________

____

10 AM ______________________

____

11 AM ______________________

____

12 NOON ______________________

____

1 PM ______________________

____

2 PM ______________________

____

3 PM ______________________

____

4 PM ______________________

____

5 PM ______________________

____

6 PM ______________________

____

Delays / Problems ______________________

____

____

Schedule Updates / Progress ______________________

____

____

Extra Work / Authorized by ______________________

____

Supervisor's Signature ______________________

JOB NAME: ______________________ DATE: __________

CONTRACTOR: ______________________ Weather: AM ______ PM ______

Expenses / Materials

Material Deliveries

Equipment Use / Hours

Equipment Rentals

Daily Work Force **No.**

Superintendent ______

Bricklayers ______

Carpenters ______

Masons ______

Electricians ______

Iron Workers ______

Plumbers ______

Others ______

__________ ______

Total ______

DAILY WORK LOG

7 AM

8 AM

9 AM

10 AM

11 AM

12 NOON

1 PM

2 PM

3 PM

4 PM

5 PM

6 PM

Delays / Problems

Schedule Updates / Progress

Extra Work / Authorized by

Supervisor's Signature ______________________

JOB NAME: ______________________ DATE: ______________

CONTRACTOR: ______________________ Weather: AM ______ PM ______

Expenses / Materials

____

Material Deliveries

____

Equipment Use / Hours

____

Equipment Rentals

____

Daily Work Force	**No.**
Superintendent	______
Bricklayers	______
Carpenters	______
Masons	______
Electricians	______
Iron Workers	______
Plumbers	______
Others	______
______________	______
Total	______

DAILY WORK LOG

7 AM ______________________

8 AM ______________________

9 AM ______________________

10 AM ______________________

11 AM ______________________

12 NOON ______________________

1 PM ______________________

2 PM ______________________

3 PM ______________________

4 PM ______________________

5 PM ______________________

6 PM ______________________

Delays / Problems

____

Schedule Updates / Progress

____

Extra Work / Authorized by

____

Supervisor's Signature ______________________________

JOB NAME: ______________________ DATE: ______________

CONTRACTOR: ______________________ Weather: AM ________ PM ________

DAILY WORK LOG

Expenses / Materials

Material Deliveries

Equipment Use / Hours

Equipment Rentals

Daily Work Force **No.**

Superintendent	______
Bricklayers	______
Carpenters	______
Masons	______
Electricians	______
Iron Workers	______
Plumbers	______
Others	______
____________	______
Total	______

7 AM

8 AM

9 AM

10 AM

11 AM

12 NOON

1 PM

2 PM

3 PM

4 PM

5 PM

6 PM

Delays / Problems

Schedule Updates / Progress

Extra Work / Authorized by

Supervisor's Signature ______________________

JOB NAME: ______________________ DATE: ______________

CONTRACTOR: ______________________ Weather: AM ______ PM ______

DAILY WORK LOG

Expenses / Materials	
	7 AM
	8 AM
Material Deliveries	9 AM
	10 AM
	11 AM
Equipment Use / Hours	
	12 NOON
	1 PM
	2 PM
Equipment Rentals	
	3 PM
	4 PM
	5 PM

Daily Work Force	No.
Superintendent	
Bricklayers	
Carpenters	
Masons	
Electricians	
Iron Workers	
Plumbers	
Others	
Total	

6 PM

Delays / Problems

Schedule Updates / Progress

Extra Work / Authorized by

Supervisor's Signature ______________________

JOB NAME: ______________________ DATE: ______________

CONTRACTOR: ______________________ Weather: AM ________ PM ________

DAILY WORK LOG

Expenses / Materials

Material Deliveries

Equipment Use / Hours

Equipment Rentals

Daily Work Force No.

Superintendent ______

Bricklayers ______

Carpenters ______

Masons ______

Electricians ______

Iron Workers ______

Plumbers ______

Others ______

______ ______

Total ______

7 AM

8 AM

9 AM

10 AM

11 AM

12 NOON

1 PM

2 PM

3 PM

4 PM

5 PM

6 PM

Delays / Problems

Schedule Updates / Progress

Extra Work / Authorized by

Supervisor's Signature ______________________

JOB NAME: ____________________ DATE: ____________

CONTRACTOR: ____________________ Weather: AM ______ PM ______

DAILY WORK LOG

Expenses / Materials

____
____
____
____
____

Material Deliveries

____
____
____
____
____

Equipment Use / Hours

____
____
____
____
____

Equipment Rentals

____
____
____
____
____

Daily Work Force	**No.**
Superintendent	______
Bricklayers	______
Carpenters	______
Masons	______
Electricians	______
Iron Workers	______
Plumbers	______
Others	______
____________	______
Total	______

7 AM ______________________________

8 AM ______________________________

9 AM ______________________________

10 AM ______________________________

11 AM ______________________________

12 NOON ______________________________

1 PM ______________________________

2 PM ______________________________

3 PM ______________________________

4 PM ______________________________

5 PM ______________________________

6 PM ______________________________

Delays / Problems ______________________________

Schedule Updates / Progress ______________________________

Extra Work / Authorized by ______________________________

Supervisor's Signature ______________________________

PROJECT NOTES

JOB NAME:

PROJECT NOTES

JOB NAME:

PROJECT NOTES

JOB NAME:

PROJECT NOTES

JOB NAME:

PROJECT NOTES

JOB NAME:

PROJECT NOTES

JOB NAME:

PROJECT NOTES

JOB NAME:

ACCIDENT REPORT

JOB NAME: ______________________________

DATE: ______________________________

Supervisor: ______________________________

Name of Injured: ______________________________

Job Classification: ____________ DOB: ____________ SSN: ____________

Description of Injury:

____

____

____

Accident Scene Detail (exact location, time of occurrence, equipment involved, etc.):

____

____

____

____

First Aid Treatment Required:

____

____

____

Employee Remarks:

____

____

____

Corrective Actions Required:

____

____

____

Supervisor's Signature: ____________ **Date:** ____________

ACCIDENT REPORT

JOB NAME: ______________________________

DATE: ______________________________

Supervisor: ______________________________

Name of Injured: ______________________________

Job Classification: ____________ DOB: ____________ SSN: ____________

Description of Injury:

____

____

____

Accident Scene Detail (exact location, time of occurrence, equipment involved, etc.):

____

____

____

____

First Aid Treatment Required:

____

____

____

Employee Remarks:

____

____

____

Corrective Actions Required:

____

____

____

Supervisor's Signature: ____________ **Date:** ____________

ACCIDENT REPORT

JOB NAME: ______________________________

DATE: ______________________________

Supervisor: ______________________________

Name of Injured: ______________________________

Job Classification: ____________________ DOB: ____________ SSN: ____________

Description of Injury:

____

____

____

Accident Scene Detail (exact location, time of occurrence, equipment involved, etc.):

____

____

____

____

First Aid Treatment Required:

____

____

____

Employee Remarks:

____

____

____

Corrective Actions Required:

____

____

____

Supervisor's Signature: ____________________ **Date:** ____________

ACCIDENT REPORT

JOB NAME: ______________________________

DATE: ______________________________

Supervisor: ______________________________

Name of Injured: ______________________________

Job Classification: ____________ DOB: ____________ SSN: ____________

Description of Injury:

____

____

____

Accident Scene Detail (exact location, time of occurrence, equipment involved, etc.):

____

____

____

____

First Aid Treatment Required:

____

____

____

Employee Remarks:

____

____

____

Corrective Actions Required:

____

____

____

Supervisor's Signature: ____________ **Date:** ____________

ACCIDENT REPORT

JOB NAME: ____________________

DATE: ____________________

Supervisor: ____________________

Name of Injured: ____________________

Job Classification: ____________________ DOB: __________ SSN: __________

Description of Injury:

Accident Scene Detail (exact location, time of occurrence, equipment involved, etc.):

First Aid Treatment Required:

Employee Remarks:

Corrective Actions Required:

Supervisor's Signature: ____________________ **Date:** __________

ACCIDENT REPORT

JOB NAME: ______________________________

DATE: ______________________________

Supervisor: ______________________________

Name of Injured: ______________________________

Job Classification: ____________ DOB: ________ SSN: ________

Description of Injury:

____

____

____

Accident Scene Detail (exact location, time of occurrence, equipment involved, etc.):

____

____

____

____

First Aid Treatment Required:

____

____

____

Employee Remarks:

____

____

____

Corrective Actions Required:

____

____

____

Supervisor's Signature: ____________________ **Date:** ____________

ACCIDENT REPORT

JOB NAME: ______________________________

DATE: ______________________________

____

Supervisor: ______________________________

Name of Injured: ______________________________

Job Classification: ______________ DOB: __________ SSN: __________

Description of Injury:

____

____

____

Accident Scene Detail (exact location, time of occurrence, equipment involved, etc.):

____

____

____

____

First Aid Treatment Required:

____

____

____

Employee Remarks:

____

____

____

Corrective Actions Required:

____

____

____

Supervisor's Signature: ______________ **Date:** __________

SAMPLE TIME AND MATERIAL FORM

Bill To: ______________________

____

____

Date ______________________

Project ______________________

Project No. ______________________

Description of Work ______________________

____

____

Location ______________________

Authorized by ______________________

Approved by ______________________

LABOR

Date	Employee	Work Performed	Hours		Rate	Amount
			REG.	OT		

MATERIALS

Date	Description	Quantity	Unit Price	Amount

EQUIPMENT

Date	Quantity	Type/size	Work Performed	Hours	Rate	Amount

TOTAL ______________________

Prepared by: ______________________

Accepted by: ______________________

Date ______________________

DAILY FIELD REPORT

Page 1 of ______

Project: ______________________ No: ________ Date: __________________

Location: ______________________________ Weather: ______________________

Superintendent: __________________________ Temp: 8AM ______ 1PM ______ 4PM ______

STAFF

Name	Classification	Name	Classification

EQUIPMENT

Quant.	Type / Size	Work / Idle	Work Performed	Arrival	Departure
		/			
		/			
		/			
		/			
		/			
		/			
		/			
		/			

VISITORS / CONVERSATIONS / MEETINGS

REQUIRED MATERIALS / INFORMATION

Item	Requested from	Company	Promised by

______________________________ ______________________________

Project Manager (Signature) Superintendent (Signature)

DAILY FIELD REPORT (continued)

Page_____ of_____

FIELD WORK REPORT

Quant.	Classification	Sub.	Description and Location of Work	Co. #
	TOTAL			

JOBSITE SAFETY REVIEW CHECKLIST

Periodically review the jobsite, considering how each of the items below has been accommodated. If treatment for any item is inadequate, notify the responsible party and ensure its immediate correction.

Condition	Date Completed
NOTIFICATION	
1. Safety signs in place	______
2. Emergency phone numbers posted	______
3. Safety meetings held	______
4. Fire extinguisher signs posted	______
5. Job safety representative designated	______
6. Fire alarm signs posted	______
7. All exits clearly marked	______
8. "No Smoking" signs posted	______
9. Evacuation plan posted	______
10. Hot work permits secured	______
11. All personnel and pedestrians notified of loud construction noises	______
12. Warnings and instructions to the public posted	______
SAFETY EQUIPMENT AND CLOTHING ON HAND	
1. Fire extinguishers (correct quantity type)	______
2. First-aid kit (filled and supplied)	______
3. Stretcher	______
4. Temporary fire alarm operating	______
5. First-aid room designated	______
6. Safety harnesses, ropes, slings (inspected)	______
7. Safety nets (inspected)	______
8. Hard hats, eye protection, gloves, safety shoes	______
9. No loose clothing or jewelry	______
JOB SITE AND PERSONNEL SAFETY SECURITY	
1. All construction vehicles properly identified	______
2. No privately owned operated vehicles on site	______
3. Fuel tanks and combustibles stored properly	______
4. Good housekeeping (no trash accumulation)	______
5. No unauthorized fires, burning, welding, etc.	______
6. Smoking prohibited in sensitive areas	______
7. Fire extinguishers and protection equipment present in hot work areas	______
TEMPORARY POWER, LIGHTING, AND SMALL TOOLS	
1. All temporary electric systems properly grounded	______
2. All extension cords of three-wire type	______
3. All tools and equipment insulated and or grounded	______
4. All hand tools in good repair	______
5. All extension cords in good condition	______
6. All work areas properly lit	______
7. All electric panels and exposed wiring inaccessible to unauthorized personnel	______
8. All extension cords and temporary power receptacles using ground fault circuit interrupters	______
9. All grounding conductors tested for continuity	______
10. Temporary power constructions closed to weather	______

Condition	Date Completed
EXCAVATIONS	
1. All shoring and earth retention systems properly designed and in place	______
2. Excavations properly dewatered	______
3. Has condition of excavations open for an extended period of time changed at all (erosion, water content)?	______
4. Are all excavations large enough to complete work without unreasonable restriction?	______
5. Excavations ventilated and free of flammable and or toxic gasses	______
PERSONNEL PROTECTION	
1. Building perimeter protected at each floor per OSHA for the respective conditions	______
2. All floor openings protected or closed off	______
3. Ladders properly set and secured	______
4. Temporary bridges supported and with rails	______
5. All traffic areas free of materials and debris	______
6. All flammable debris removed daily	
7. Adequate temporary lighting in all areas	______
8. Temporary heaters properly located	______
9. All materials and debris away from temp heaters	______
10. Oxygen, acetylene, and other fuel tanks properly stored and secured	______
11. Confined spaces properly ventilated	______
12. All drinking water potable	______
13. All work areas sanitary	
PEDESTRIAN PUBLIC PROTECTION	
1. Entire site protected from unauthorized entry	______
2. Dangerous areas within the site restricted from nonconstruction personnel	______
3. All sidewalk sheds, barricades, overhead protection, and warning lights in place	______
4. Temporary pedestrian traffic areas:	
a. Properly illuminated	______
b. Free of trip hazards, debris, materials, sharp objects	______
5. All construction noises held to reasonable levels	______
6. Work performed in areas occupied by the public properly authorized	______
7. Appropriate warnings and instructions posted	______
LIABILITY	
1. All release forms executed and delivered for trades to use staging, hoists, elevators, and equipment	______
2. Arrange for OSHA safety inspection to advise of additional recommended precautions	______
3. Arrange for job site inspection by liability insurance carrier. Either secure a favorable written report, or immediately make all recommended corrections and reinspect.	______

WINTER CONDITIONS CHECKLIST

This checklist is a summary of key items to consider as each construction project approaches the winter period.

PROJECT STATUS	YES	NO
1. Building portions satisfactorily closed to the weather:		
a. Roof and flashings	___	___
b. Doors and windows	___	___
c. Building skin	___	___
d. ____________	___	___
2. Permanent building systems usable for temporary heat:		
a. Electrical	___	___
b. HVAC	___	___
3. All open excavations closed prior to freezing conditions	___	___
4. Permanent sources of power and fuel available	___	___
5. Temporary power provisions necessary	___	___
6. Temporary fuel provisions necessary	___	___

CONTRACT	YES	NO
1. Who is responsible for temporary heat and protection?		
a. The owner	___	___
b. Prime contractor or construction manager	___	___
c. Sub or trade contractor	___	___
2. Temporary heat required between: ____________ (date) and ____________ (date)		
3. Temporary heat and protection now required because of a delay	___	___
4. If (3) is yes, who is responsible?		
a. The owner	___	___
b. Owner's agents	___	___
c. Prime contractor or construction manager	___	___
d. Sub or trade contractor (Name) ____________	___	___
5. Party named in (4) notified in writing	___	___
6. Party named in (4) accepted responsibility	___	___
7. If (6) is no, backcharge procedure has begun	___	___
8. Estimated cost of temporary		
a. Protection $ ________		
b. Heat $ ________		
c. Light and power $ ________		

JOB PRECAUTIONS	YES	NO
1. Arrangements have been made to secure:		
a. Temporary protection materials	___	___
b. Temporary enclosure materials	___	___
c. Temporary heaters	___	___
d. Continuous fuel supply	___	___
2. Temporary heaters are:		
a. Of adequate size and type	___	___
b. Fully operational and maintained	___	___
c. Of type allowed by codes	___	___
d. Situated in a safe manner relative to personnel, pedestrians, traffic, building materials, and ventilation	___	___
e. On a service/maintenance schedule	___	___
3. Temporary fuel is:		
a. On hand in adequate supply	___	___
b. Properly and safely stored	___	___
c. On a set refueling schedule	___	___
4. All water pockets have been eliminated:		
a. Roof areas	___	___
b. Pavement and graded areas	___	___
c. Sleeves, inserts, chases, and openings	___	___
e. Other ____________	___	___
5. Arrangements have been made for:		
a. Snow plowing and removal	___	___
b. Equipment cold weather protection	___	___
c. Vehicle maintenance	___	___
6. Precautions have been taken to protect exposed work:		
a. Exposed piping protected, drained or heat traced	___	___
b. Recently placed work (concrete, formwork reinforcing steel, masonry, etc.)	___	___
7. All project areas have been adequately marked to avoid damage during snow removal:		
a. Parking areas	___	___
b. Entrances, exits, gates, passageways	___	___
c. Pedestrian traffic areas	___	___
d. Material and fuel storage areas	___	___
8. Any necessary photographs of all pre-winter jobsite conditions have been taken for a record	___	___

	YES	NO
HAS THIS CHECKLIST BEEN SENT TO ALL JOB SITES?	___	___

DELAY DAMAGES CHECKLIST: POTENTIAL CATEGORIES OF DAMAGES

Note: This checklist includes items to be considered when assessing costs resulting from a job delay. This list is intended for general discussion purposes only. In order to determine your actual potential claims in any given situation, it is recommended that you seek legal counsel.

- increased labor costs
  - increased hourly wages and benefits
  - increased hours due to out-of-sequence work; acceleration; disruption
- increased material costs
- increased equipment costs
  - increased rates for equipment and operators
  - increased hours due to out-of-sequence work; acceleration; disruption
  - idle equipment costs
  - mobilization costs
- increased direct overhead costs
  - increased supervision costs at site
  - trailer(s) and utilities
  - temporary heat and light
  - phones
  - sanitation facilities
  - clean-up
  - bond/insurance
- increased home office overhead costs
  - allocated home office overhead costs [or]
  - direct administrative personnel costs
    - based on allocated hourly rates
- lost profits
- loss of bonding capacity
  - lost profits on work which could otherwise have been obtained and performed but for loss of bonding capacity and delays
- interest
- loss of use of money
- attorney's fees
- punitive damages

TIME MANAGEMENT TIPS

Time management is two things. First is increasing efficiency; refining your own management techniques to increase output for a given period of time. Second is learning to work smarter. Allow time for other activities that you enjoy doing. This reduces stress on your system and, in turn, improves your operating efficiency.

1. Develop observation skills. Observe intently what is going on around you. Improving ability to get clear, accurate impressions increases the odds of correct initial responses.
2. Improve capacity for observation and quick decision making by increasing alertness, energy level, knowledge base, and experience.
3. Increase your alertness by
   - Overcoming natural tendencies to become preoccupied.
   - Changing routines.
   - Practicing daily a relevant skill that interests you.
   - Cultivating interests centering on observation.
4. Improve energy levels by
   - Eliminating personal criticisms, defensiveness, and other negative effects that drain energy and attention.
   - Becoming aware of those times that you lose energy by establishing times during the day to check on your activities on all levels.
   - Establishing and maintaining an exercise program to improve levels of overall physical fitness.
   - Using the creative power of sleep. The more demands you make on yourself, the more sleep you will need.
5. Give attention every day to expanding your knowledge and experience, and increasing your managerial skills.
6. When you can't find an answer, stop. Let the problem cool. Save time by restating the problem and observing it from a different angle.
7. Talking is more than transmitting words. You speak with your whole organism. Hearing your own words as you explain your problem to another often leads you to flow directly toward the answer.
8. Use language with precision. Avoid the possibility of confusion resulting from extraneous details.
9. Be sure that you understand statements by others by restating the concept in your own words to get a "yes" response.
10. Take the responsibility to be sure that others completely understand you before proceeding.
11. Draw diagrams to get more understanding and agreement in less time.
12. Remember that one appropriate analogy is often worth more than hours of discussion.
13. It is not as important to be able to read rapidly as it is to be able to decide what not to read.
14. Set priorities. Decide what are the most important activities and arrange your efforts specifically around them.
15. Organize your day. Have a definite game plan based upon your priorities. Control interruptions. Don't let the "immediate" demands interfere with your plan.
16. Become "now" oriented. Once you decide on an activity, *focus* your energy on it until it is completed or filed for future reference.
17. Delegate. Develop the skills to train others, then depend on them. Delegate as much as you possibly can.
18. Start with the tough jobs. Do the most important work early, when your energy levels are at their highest. Save busy work and errands for later, lower energy periods.
19. Reduce meetings. Resolve as much as you can by phone. Send subordinates wherever possible. Schedule meetings to run up against the noon hour or day's end to cut ramblings.
20. Avoid procrastination. The pressure of deadlines creates inefficiency, ineffectiveness, and rework.

JOB START-UP CHECKLIST

Use this checklist to arrange for all services, supplies, and facilities necessary for a smooth and complete transition into your project's construction phase.

CONTRACT

1. Type (GC, CM, CM w/GMP, etc.) ______________
2. Contract signed (date) ______________
3. Start date ______________
4. Completion date ______________
5. Number of working days ______________
6. Liquidated damages: Yes: $ ______________ /Day
   No: ______
7. Unusual restrictions: ______________
8. Other ______________

CONTRACT EXECUTION

1. Permits obtained:

Item	Who Pays	Received
a. General building permit	______	______
b. Plumbing	______	______
c. HVAC	______	______
d. Fire protection	______	______
e. Electrical	______	______
f. Other: ______	______	______

2. Billing procedure:
   a. Date subcontractor requisitions due: ______________
   b. Date general requisition to owner due: ______________
   c. Schedule of values prepared/approved ______________
   d. Remarks ______________
3. Change order procedure:
   a. Change clause present: Section ______________
   b. Forms required: ______________
   c. Remarks ______________
4. EEO Requirements:
   a. Mandatory ______________ b. Good faith ______________
5. Independent testing laboratories:
   a. Areas required: ______________
   b. Payment responsibility: ______________
6. Baselines and benchmark:
   Responsibility: ______________
7. Job meeting schedule: ______________
8. Dispute resolution:
   a. Dispute clause present: Section ______________
   b. Arbitration provision: Section ______________
   c. Notice requirements: ______________

CONTRACT DOCUMENTS

1. Jobsite copies of:
   a. Contract ______
   b. General, special, and supplementary conditions ______
   c. Technical specifications ______
   d. Plans ______
   e. Project manual/procedures ______

COST AND PRODUCTION CONTROL

1. Documents on file:
   a. Budget ______
   b. Resource estimates: labor ______
      materials ______
   c. Job cost report ______
   d. Other ______________ ______

SITE AND SERVICES

1. Temporary fences, protection (see safety checklist) ______
2. Guard service ______
3. Temporary electric ______
4. Temporary water ______
5. Dumpster, disposal arrangements ______
6. Progress photograph service ______
7. Testing laboratories
   a. Soils ______
   b. Concrete ______
   c. Steel and welding ______
   d. Other ______________ ______
8. Weather information phone numbers

ADMINISTRATION

1. Supply of job forms:
   a. Field reports ______
   b. Change order forms ______
   c. Change order summary logs ______
   d. Quotation and telephone quotation forms ______
   e. Payroll forms ______
   f. Time and material tickets ______
   g. Subcontract adjustment forms ______
   h. Job meeting forms ______
   I. Memos ______
   j. Schedule status report forms ______
   k. Cost report forms ______
   l. Photograph record forms ______
   m. Other ______________ ______
   n. Other ______________ ______
2. Start-up submissions on file:
   a. Subcontractor payment and performance bonds ______
   b. Subcontractor insurance certificates ______
   c. Equipment use releases ______
   d. Shop drawing submission schedule ______
   e. Other ______________ ______
3. Project files
   a. Contract and correspondence files ______
   b. Submittal files ______
   c. Special files ______

FIELD OFFICE AND OFFICE EQUIPMENT

1. Trailer(s) ______
2. Retail space ______
3. Temporary facilities
   a. Heat ______
   b. Lighting and power ______
   c. Telephone(s) and mobile phones ______
   d. Site radios ______
   e. Lavatories ______
   f. Water ______
4. Office furniture
   a. Desks ______
   b. Conference table(s) ______
   c. Plan table(s) ______
   d. Swivel chairs, folding chairs, stools ______
   e. File cabinets, fireproof file cabinet ______
   f. Plan rack(s), plan edge reinforcing machine ______
   g. Bookcases, tack boards ______
5. Job directory
   a. Owner representative(s) ______
   b. Design professionals ______
   c. Government and approving authorities ______
   d. Police, fire, hospital, security ______
   e. Jobsite personnel home numbers ______
   f. Subcontractors and suppliers ______
6. Fire and intrusion alarm systems ______
7. Safety equipment
   a. Fire extinguishers ______
   b. Hard hats
   c. First-aid kit and supplies ______
   d. Emergency phone numbers ______
   e. Stretcher ______
8. Office equipment and supplies
   a. Copier, supplies and maintenance arrangements ______
   b. Blueprint arrangements ______
   c. Computer(s), printer(s), plotter(s), software ______
   d. Telecopier ______
   e. Refrigerator, coffee machine, supplies ______
   f. Bottled water ______

PROJECT CLOSEOUT CHECKLIST

SUBCONTRACTOR SUBMITTALS

1. Each subcontractor has delivered as required:
   a. As-built drawings ________
   b. Guarantees and warrantees ________
   c. Inspection certificates ________
   d. Material/installation certifications ________
   e. Operating and maintenance manuals ________
   f. Operating instruction to owner personnel performed ________
   g. General releases ________
   h. Lien waivers ________
   i. Other ________________________ ________
   j. Other ________________________ ________

FINAL SUBMITTALS TO THE OWNER

1. As-built plans and specifications ________
2. Guarantee(s), transfer of subcontractors' guarantees ________
3. Transfer of all certifications, releases, lien waivers, operating and maintenance manuals ________
4. Other ________________________ ________
5. Other ________________________ ________

FINAL COMPLETION OF THE WORK

1. Certificate(s) of occupancy received ________
2. Punch list confirmed to be complete ________
3. Final/finish cleanup performed ________
4. Demobilization of all field facilities and equipment complete ________
5. Termination of temporary services complete:
   a. Heat, light, power, and telephone ________
   b. Fire, police, guard service ________
   c. Insurance ________
   d. Office equipment and furnishings ________
6. Owner/architect certificates of completion received ________
7. Other ________________________ ________
8. Other ________________________ ________

BILLINGS, CHARGES, AND PAYMENTS

1. All owner-acknowledged change orders submitted and approved ________
2. All subcontractor changes and adjustments processed ________
3. Steps taken toward resolution of outstanding subcontractor claims ________
4. Steps taken toward resolution of outstanding claims to the owner ________
5. All subcontractor backcharges finalized ________
6. Final billings received from all subcontractors and suppliers in acceptable form ________
7. Final billing including retainage release submitted ________

REVIEW THE CONTRACT, SUBCONTRACT OR PURCHASE ORDER, SPECIFICATIONS, AND COMPANY PROCEDURE TO DETERMINE ANY ADDITIONAL REQUIREMENTS. LIST BELOW:

1. ________________________ ________
2. ________________________ ________
3. ________________________ ________
4. ________________________ ________
5. ________________________ ________

SUBCONTRACTOR/SUPPLIER PROGRESS PAYMENT CHECKLIST

REQUIRED SUBMISSIONS

1. Contract or purchase order executed and on file ________
2. Payment and performance bond received ________
3. Insurance certificate with adequate levels of coverage received ________
4. Lien waivers received ________
5. Material certifications received ________
6. "Passing" inspection reports received ________
7. Equipment, scaffolding, or elevator use release forms received ________
8. Other ____ ________

PAYMENT CONDITIONS

1. All material billed properly approved ________
2. All material requisitioned for on-site ________
3. If material billed for is off-site:
   a. Does the contract allow for payment? ________
   b. If so, are all required insurances, title transfers, and other requirements complete? ________
   c. Does the material require inspection? ________
   d. Will the party pay for the inspection? ________

THE REQUISITIONS FOR PAYMENT

1. Submitted on time ________
2. Each line item approved by the proper authority:
   a. Contract work ________
   b. Change order work ________
3. All supporting documentation verified ________
4. All supporting time and material tickets approved ________
5. Proper retainage percentages deducted ________
6. Payment *for this work* received from the owner
7. If (6) is no, can full or partial payment still be released? ________

REVIEW THE CONTRACT, SUBCONTRACT OR PURCHASE ORDER, SPECIFICATIONS, AND COMPANY PROCEDURE TO DETERMINE ANY ADDITIONAL REQUIREMENTS. LIST BELOW:

1. ____
2. ____
3. ____

SUBCONTRACTOR/SUPPLIER FINAL PAYMENT CHECKLIST

THE WORK

1. Punch list complete ________
2. *Written* acceptances for all work received from:
   a. The owner ________
   b. The design professionals ________
   c. Our company ________
3. Demobilization complete ________
4. All temporary construction and facilities removed ________
5. Other ________________________________ ________

FINAL DOCUMENT SUBMISSIONS

1. As-built drawings ________
2. Guarantees and warrantees ________
3. Operating and maintenance manuals ________
4. Operating and maintenance instruction performed ________
5. Material certifications received ________
6. "Passing" testing and inspection reports received ________
7. Lien waivers received ________
8. Other ________________________________ ________

FINAL PAYMENT

1. All material billed for properly submitted approved ________
2. Each requisition line item approved ________
3. All supporting documentation verified ________
4. All supporting time and material tickets approved ________
5. All charge orders approved in the submitted amounts ________
6. All back charges resolved to your satisfaction ________
7. Outstanding claims by
   a. The subcontractor ________
   b. The owner ________
   c. Our company ________
8. Final payment including retainage received from the owner ________

REVIEW THE CONTRACT, SUBCONTRACT OR PURCHASE ORDER, SPECIFICATIONS, AND COMPANY PROCEDURE TO DETERMINE ANY ADDITIONAL REQUIREMENTS. LIST BELOW:

1. ________________________________
2. ________________________________
3. ________________________________

JOB MEETING GUIDELINES

Use these guidelines as an aid to prepare and follow through on all project meetings.

GENERAL

1. Job meetings and their minutes are critically important to job coordination, documentation, and protection of your interests. If properly completed, the job meeting minutes will:
   a. Record history of all significant (or potentially significant) events.
   b. Keep open items on the front burner until they are finally resolved or filed for future reference.
   c. Force action.
   d. Clarify accountability.
   e. Provide a basis to identify required expedited action.
   f. Support interpretations and serious actions.
   g. Facilitate fast, efficient research, both soon and long after the occurrence of an event.

MEETING PREPARATION

1. Schedule morning meetings if possible.
2. Always start the meetings precisely *on time,* regardless of who may be absent.
3. Make meeting attendance *mandatory.* Absence is no excuse. All parties are responsible for information contained in the meeting.
4. Confirm attendance by all those required to attend *prior* to the meeting.
5. Notify absent parties *immediately after the meeting* of decisions, consequences, and so on that affect them.

MEETING MINUTES GUIDELINES

1. Use outline format.
   a. Consecutively number meetings.
   b. Separate "Old" and "New" business.
   c. Number each successive item for easy reference.
2. Use a title for each item.
   a. Summary description of the issue.
   b. Use exactly the same title wording at each meeting.
3. Include all appropriate file references in an item title.
   a. Change order number
   b. Estimate number
   c. Architect's bulletin number
4. Name names. Pin down direct responsibility.
5. Use *short* but specific statements.
6. Read back noted language to confirm agreement on its accuracy.
7. Require definite action. Do not leave an item without determining:
   a. The next step
   b. Who is to perform it
   c. The precise date action is required by.
8. Notify all meeting recipients to immediately notify the writer of any errors or omissions in the representations.

MECHANICAL PROPERTIES OF MATERIALS

Material	Designation Number	Description	Tensile ultimate strength 10^3 psi	Tensile yield strength 10^3 psi	Shear strength 10^3 psi	Compressive strength 10^3 psi	Endurance limit 10^3 psi	Modulus of Elasticity in tension 10^6 psi	Elongation in 2″, %
Cast iron	ASTM 20		20		32	95	10	12	
	ASTM 30		30		44	115	14.5	15	
	ASTM 40		40		57	143	21	17	
	ASTM 60		60			170		19.5	
Malleable Cast iron	ASTM 47-33	Grade 32510	50	32				25	10
Meehanite			55	35					10
Carbon Steels	AISI-C1015	Hot-rolled	60	35				30	35
		Cold-drawn	75	65				30	20
	AISI-C1020	Hot-rolled	65	43				30	35
		Cold-drawn	78	66				30	20
	AISI-C1030	Hot-rolled	80	47				30	30
		Cold-drawn	93	78				30	17
	AISI-C1040	Hot-rolled	84	53				30	28
		Cold-drawn	100	92				30	15
Stainless Steel	302 (18-8)	Annealed	90	35			42	28	55
		cold-drawn bars	115	85				28	24
Aluminum Alloys	24S-O	Annealed	26	11	18		12	10.6	19
	24S-T4	Heat-treated	64	42	41		18	10.6	19
	113	As cast	24	15	20		10		1.5
	195-T4	Heat-treated casting	32	16	24		6		8.5
Magnesium Alloy	A 10	As sand-cast	22	12	18		10	6.5	2
		Heat-treated & aged casting	40	19	22		10	6.5	1
		Extruded bar	51	38	23		18	6.5	9
Bronze	Leaded tin	As cast	38	16	35			13	35

WEIGHT OF EARTH MATERIALS IN THE GROUND

Material	Weight		To convert from cu. yds. to tons, multiply by
	Lbs. per cu. ft.	Lbs. per cu. yd.	
Ashes or cinders	45	1,215	0.608
Slag	70	1,890	0.945
Earth, compact	95	2,565	1.283
Earth, dry & loose	75	2,025	1.013
Earth, moist	80	2,160	1.080
Clay, very dry	75	2,025	1.013
Clay, dry	85	2,295	1.148
Clay, moist	95	2,565	1.283
Clay, wet	110	2,970	1.485
Clay & gravel mixed, dry	100	2,700	1.350
Crushed stone	95	2,565	1.283
Sand, loose & dry	95	2,565	1.283
Sand, moist	105	2,835	1.418
Sand, wet	115	3,105	1.553
Sand & gravel, dry	100	2,700	1.350
Sand & gravel, compact	110	2,970	1.485
Sand & gravel, wet	120	3,240	1.620
Gravel, loose & dry	100	2,700	1.350
Gravel, compact	115	3,105	1.553
Gravel, wet	125	3,375	1.688
Mud, flowing	110	2,970	1.485
Mud, compact	120	3,240	1.620
Sandstone	150	4,050	2.025
Limestone	160	4,320	2.160
Granite	165	4,455	2.228

These weights are approximate, being average figures: the weight of earth material varies according to its density and moisture content.

WEIGHTS OF BUILDING MATERIALS

Material	Weights, lbs. per cu. ft.
Concrete, cinder	90–110
Concrete, slag	120–130
Concrete, stone, not reinforced	144
Concrete, stone, reinforced	150
Brickwork in place, medium to soft bricks	110
Brickwork in place, hard bricks	120–140
Sandstone masonry in place	130
Bluestone masonry in place	135
Limestone masonry in place	150
Marble in place	155
Granite in place	155
Lumber, softwood	30–33
Lumber, hardwood	35–45

APPROXIMATE SHOVEL DIGGING-LOADING CYCLES (IN SECONDS) FOR VARIOUS DEGREES OF SWING

Size of Shovel Dipper	Easy Digging (Moist Loam or Light Sandy Clay) Degree of Swing				Medium Digging (Good Common Earth) Degree of Swing				Hard Digging (Hard Tough Clay) Degree of Swing			
	45	90	135	180	45	90	135	180	45	90	135	180
3/8	12	16	19	22	15	19	23	26	19	24	29	33
1/2	12	16	19	22	15	19	23	26	19	24	29	33
3/4	13	17	20	23	16	20	24	27	20	25	30	34
1	14	18	21	25	17	21	25	29	21	26	31	36
1 1/4	14	18	21	25	17	21	25	29	21	26	31	36
1 1/2	15	19	23	27	18	23	27	31	22	28	33	38
1 3/4	16	20	24	28	19	24	28	32	23	29	34	39
2	17	21	25	30	20	25	29	34	24	30	35	41
2 1/2	18	22	27	32	21	26	31	36	25	31	37	43

(Based on no delays, digging in optimum depths of cut, loading trucks on same grade as machine)

STOCKPILE PRODUCTION WITH RUBBER TIRED FRONT END LOADERS

Cycle Distance (Ft.)	Bucket Capacity (Cubic Yards) 1	1 1/2	2	2 1/2	3 1/2	4 1/2	6 1/2	12
50	175	260	350	460	500	660	850	1500
100	130	200	250	340	390	520	700	1175
150	100	155	180	260	320	430	610	1000
200	85	125	150	210	280	380	550	875
250	70	105	120	180	250	340	500	825
300	60	95	105	165	230	320	450	750
350	57	85	90	150	220	290	410	700
400	55	80	85	140	205	270	385	650

(In tons per hour. Based on 3000 lb./cu. yd. material, level firm ground and theoretical 60-minute hour)

POWER SHOVEL HOURLY OUTPUT (CUBIC YARDS)

Class of Material	Size of Dipper (Yd.) 3/8	1/2	3/4	1	1 1/4	1 1/2	1 3/4	2	2 1/2
Moist Loam or Light Sand Clay	85	115	165	205	250	280	320	355	405
Sand and Gravel	80	110	155	200	230	270	300	330	390
Good Common Earth	70	95	135	175	210	240	270	300	350
Clay-Hard, Tough	50	75	110	145	180	210	235	265	310
Rock-Well Blasted	40	60	95	125	155	180	205	230	275
Common Earth with Rocks and Roots	30	50	80	105	130	155	180	200	245
Clay—Wet and Sticky	25	40	70	95	120	145	165	185	230
Rock—Poorly Blasted	15	25	50	75	95	115	140	160	195

(Based on optimum depth of cut 90° swing and grade level loading)

WEIGHTS OF MISCELLANEOUS MATERIALS

Description	Weight Pounds Per Cu. Ft.
METALS, ALLOYS, ORES	
aluminum, cast-hammered	165
aluminum, bronze	481
antimony	416
brass, cast-rolled	534
bronze, 7.9 to 14% Sn	509
chromium	428
copper, cast-rolled	556
copper, ore, pyrites	262
gold, cast-hammered	1205
iron, cast, pig	450
iron, wrought	485
iron, steel	490
iron, spiegel-eisen	468
iron, ferro-silicon	437
iron ore, hematite	325
iron ore, hematite in	160
bank	to 180
loose	130
	to 160
iron ore, limonite	237
iron ore, magnetite	315
iron slag	172
lead	706
lead ore, galena	465
magnesium	109
manganese	456
mercury	848
molybdenum	562
nickel	545
nickel, monel metal	556
platinum, cast-hammered	1330
silver, cast-hammered	656
tin, cast-hammered	459
tin, babbit metal	443
tin ore, cassiterite	418
tungsten	1180
vanadium	350
zinc, cast-rolled	440
zinc ore, blende	253
MINERALS	
asbestos	153
barytes	281
basalt	184
bauxite	159
borax	109
chalk	137
clay, marl	137
dolomite	181
feldspar, orthoclase	159
gneiss, serpentine	159
granite, syenite	175
greenstone, trap	187
gypsum, alabaster	159
hornblende	187
limestone, marble	165

Description	Weight Pounds Per Cu. Ft.
magnesite	187
phosphate rock, apatite	200
porphyry	172
pumice, natural	40
quartz, flint	165
sandstone, bluestone	147
shale, slate	175
soapstone, talc	169
BITUMINOUS MATERIALS	
asphaltum	81
coal, anthracite	97
coal, bituminous	84
coal, lignite	78
coal, peat, turf, dry	47
coal, charcoal, pine	23
coal, charcoal, oak	33
coal, coke	75
graphite	131
paraffin	56
petroleum, crude	55
petroleum, refined	50
petroleum, benzine	46
petroleum, gasoline	42
pitch	69
tar, bituminous	75
TIMBER, U.S. SEASONED	
ash, white-red	40
cedar, white-red	22
chestnut	41
cypress	30
fir, douglas spruce	32
fir, eastern	25
elm, white	45
hemlock	29
hickory	49
locust	46
maple, hard	43
maple, white	33
oak, chestnut	54
oak, live	59
oak, red, black	41
oak, white	46
pine, oregon	32
pine, red	30
pine, white	26
pine, yellow, long leaf	44
pine, yellow, short leaf	38
poplar	30
redwood, california	26
spruce, white, black	27
walnut, black	38
walnut, white	26

Moisture Contents
Seasoned timber 15 to 20%
Green timber up to 50%

Description		Weight Pounds Per Cu. Ft.
VARIOUS SOLIDS		
carbon, amorphous, graphitic		129
cork		15
ebony		76
fats		58
glass, common, plate		160
glass, crystal		184
glass, flint		220
phosphorus, white		114
porcelain, china		150
resins, rosin, amber		67
rubber, caoutchouc		58
silicon		155
sulphur, amorphous		128
wax		60
VARIOUS LIQUIDS		
alcohol,	100%	49
acids, muriatic	40%	75
acids, nitric	91%	94
acids, sulphuric	87%	112
lye, soda	66%	106
oils, vegetable		58
oils, mineral, lubricants		57
petroleum		55
gasoline		42
water, 4 deg. C., max. density		62.428
100 deg. C.		59.830
ice		56
snow, fresh fallen		8
sea water		64
EARTH—EXCAVATED		
clay, dry		63
clay, damp, plastic		110
clay & gravel, dry		110
earth, dry, loose		76
earth, dry, packed		95
earth, moist, loose		78
earth, moist, packed		96
earth, mud, flowing		108
earth, mud, packed		115
sand, gravel, dry, loose		90 to 105
sand, gravel, dry, packed		100 to 120
sand, gravel, wet		118 to 120

REINFORCING BARS
Nominal Dimensions in Inches

Bar No.	Diameter in Inches	Area Sq. In.	Perimeter in Inches	Weight per Ft. in Lbs.
3	0.375	0.11	1.178	0.376
4	0.500	0.20	1.571	0.668
5	0.625	0.31	1.963	1.043
6	0.750	0.44	2.356	1.502
7	0.875	0.60	2.749	2.044
8	1.000	0.79	3.142	2.670
9	1.128	1.00	3.544	3.400
10	1.270	1.27	3.990	4.303
11	1.410	1.56	4.430	5.313
14	1.693	2.25	5.320	7.650
18	2.257	4.00	7.090	13.600

WIRE FABRIC—ELECTRIC WELDED
Two-Way Reinforcement (In Rolls)

Spacing of Wires in Inches			Gauge of Wires		Section Areas Per Foot of Width		Approx. Weight Per 100 Sq. Ft. in Lbs.
Long.		Trans.	Long.	Trans.	Long. Sq. In.	Trans. Sq. In.	
6	x	6	10	10	.029	.029	21
6	x	6	8	8	.041	.041	30
6	x	6	6	6	.058	.058	42
6	x	6	4	4	.080	.080	58
4	x	4	10	10	.043	.043	31
4	x	4	8	8	.062	.062	44
4	x	4	6	6	.087	.087	62
4	x	4	4	4	.120	.120	85

WIRE FABRIC—ELECTRIC WELDED
Two-Way Reinforcement (In Sheets)

Spacing of Wires in Inches			Gauge of Wires		Section Areas Per Foot of Width		Approx. Weight Per 100 Sq. Ft. in Lbs.
Long.		Trans.	Long.	Trans.	Long. Sq. In.	Trans. Sq. In.	
6	x	12	00	4	.172	.040	78
6	x	6	10	10	.029	.029	21
6	x	6	6	6	.058	.058	42
6	x	6	4	4	.080	.080	58
6	x	6	2	2	.108	.108	78
6	x	6	0	0	.148	.148	107
4	x	4	4	4	.120	.120	85
4	x	4	3	3	.140	.140	100

WIRE ROPE (6 strand, 19 wires)

Diameter, inches	Approximate weight per 100 ft. lb.	Breaking Strength, Lb.		
		Mild plow steel	Plow steel	Improved plow steel
1/4	10	4,140	4,780	5,480
3/8	23	10,000	11,000	12,600
1/2	40	17,000	18,800	21,600
5/8	63	26,200	28,800	33,200
3/4	90	37,400	41,200	47,400
7/8	123	50,800	56,000	64,400
1	160	66,000	73,000	84,000
1 1/8	203	83,000	92,000	106,000
1 1/4	250	102,000	113,000	130,000
1 1/2	360	145,000	161,000	185,000

SAFE CAPACITIES OF CHAINS

Diameter of iron, inches	Safe Capacity, Lbs.			
	Used straight	Used at 60-degree angle	Used at 45-degree angle	Used at 30-degree angle
1/4	1,330	1,000	850	600
3/8	2,660	2,050	1,700	1,200
1/2	5,330	4,100	3,400	2,400
5/8	8,330	6,800	5,600	4,000
3/4	12,000	9,400	7,800	5,500
7/8	16,330	12,800	10,400	7,400
1	20,830	16,000	13,000	9,400

U.S. GALLONS IN ROUND TANKS
For One Foot in Depth

Dia. of tank Ft.	In.	No. U.S. gals.	Cu. Ft.
1		5.87	.785
1	1	6.89	.922
1	2	8	1.069
1	3	9.18	1.227
1	4	10.44	1.396
1	5	11.79	1.576
1	6	13.22	1.767
1	7	14.73	1.969
1	8	16.32	2.182
1	9	17.99	2.405
1	10	19.75	2.640
1	11	21.58	2.885
2		23.50	3.142
2	1	25.50	3.409
2	2	27.58	3.687
2	3	29.74	3.976
2	4	31.99	4.276
2	5	34.31	4.587
2	6	36.72	4.909
2	7	39.21	5.241
2	8	41.78	5.585
2	9	44.43	5.940
2	10	47.16	6.305
2	11	49.98	6.681
3		52.88	7.069
3	1	55.86	7.467
3	2	58.92	7.876
3	2	58.92	7.876
3	4	65.28	8.727
3	5	68.58	9.168
3	6	71.97	9.62
3	7	75.44	10.085
3	8	78.99	10.559
3	9	82.62	11.045
3	10	86.33	11.541
3	11	90.13	12.048
4		94	12.566
4	1	97.96	13.095
4	2	102	13.635
4	3	106.12	14.186
4	4	110.32	14.748
4	5	114.61	15.321
4	6	118.97	15.90
4	7	123.42	16.50
4	8	127.95	17.10
4	9	132.56	17.72
4	10	137.25	18.35
4	11	142.02	18.99
5		146.88	19.63
5	1	151.82	20.29
5	2	156.83	20.97
5	3	161.93	21.65
5	4	167.12	22.34
5	5	172.38	23.04
5	6	177.72	23.76
5	7	183.15	24.48
5	8	188.66	25.22
5	9	194.25	25.97
5	10	199.92	26.73
5	11	205.67	27.49
6		211.51	28.27
6	3	229.50	30.68
6	6	248.23	33.18
6	9	267.69	35.78
7		287.88	38.48
7	3	308.81	41.28
7	6	330.48	44.18
7	9	352.88	47.17
8		376.01	50.27
8	3	399.88	53.46
8	6	424.48	56.75
8	9	449.82	60.13
9		475.89	63.62
9	3	502.70	67.20
9	6	530.24	70.88
9	9	558.51	74.66
10		587.52	78.54

LUMBER CONVERTED TO BOARD FOOT MEASURE

Lumber size inches	Board Ft. per L.F.	Board Feet Per Length						
		8 FT.	10 FT.	12 FT.	14 FT.	16 FT.	18 FT.	20 FT.
1x2	0.166	1.33	1.67	2.00	2.33	2.67	3.00	3.33
1x3	0.250	2.00	2.50	3.00	3.50	4.00	4.50	5.00
1x4; 2x2	0.333	2.67	3.33	4.00	4.67	5.33	6.00	6.67
1x6; 2x3	0.500	4.00	5.00	6.00	7.00	8.00	9.00	10.00
1x8; 2x4	0.667	5.34	6.67	8.00	9.34	10.67	12.00	13.34
1x12; 2x6; 3x4	1.000	8.00	10.00	12.00	14.00	16.00	18.00	20.00
2x8; 4x4	1.333	10.66	13.33	16.00	18.66	21.33	24.00	26.66
2x10	1.667	13.34	16.67	20.00	23.34	26.67	30.00	33.34
2x12; 3x8; 4x6	2.000	16.00	20.00	24.00	28.00	32.00	36.00	40.00
2x14	2.333	18.66	23.33	28.00	32.66	37.33	42.00	46.66
3x10	2.500	20.00	25.00	30.00	35.00	40.00	45.00	50.00
4x10	3.333	26.66	33.33	40.00	46.62	53.33	60.00	66.66
3x6	1.500	12.00	15.00	18.00	21.00	24.00	27.00	30.00
3x12; 6x6	3.00	24.00	30.00	36.00	42.00	48.00	54.00	60.00
3x14	3.500	28.00	35.00	42.00	49.00	56.00	63.00	70.00
3x16; 4x12; 6x8	4.000	32.00	40.00	48.00	56.00	64.00	72.00	80.00
4x8	2.667	21.34	26.67	32.00	37.34	42.67	48.00	53.34
4x14	4.667	37.34	46.67	56.00	65.34	74.67	84.00	93.34
4x16; 8x8	5.333	42.66	53.33	64.00	74.66	85.33	96.00	106.66
6x10	5.000	40.00	50.00	60.00	70.00	80.00	90.00	100.00
8x10	6.667	53.34	66.67	80.00	93.34	106.67	120.00	133.34

One board foot = one lineal foot of 1″ x 12″ board. Therefore, at one lineal foot, a 2″ x 4″ would equal 2x4 divided by 12 or 0.667 board feet.

NAILS

Size	Length, inches	TYPE		
		Common	Casing	Finishing
		Approximate number per lb.		
2d	1	830		1.351
3d	1¼	528		807
4d	1½	316	473	584
5d	1¾	271		500
6d	2	168	236	309
7d	2¼	150	210	
8d	2½	106	145	189
9d	2¾	96		
10d	3	69	94	121
12d	3¼	63		
16d	3½	49	71	90
20d	4	31	52	62
30d	4½	24		
40d	5	18		
50d	5½	14		
60d	6	11		

UNITS REQUIRED—VARIOUS I.D. DIMENSIONS

5 Foot I.D. Bottom 14 Barrels Per Course Double Course = 30½ Barrels

Inside Diameter	No Batters Required			
Top	#1	#2	#3	#4
54″	14			
48″	14	14		
42″	14	14	12	
36″	14	14	12	12
30″	14	14	12	22
24″	14	14	12	32

UNITS REQUIRED—VARIOUS I.D. DIMENSIONS

6 Foot I.D. Bottom 16½ Barrels Per Course Double Course = 35½ Barrels

Inside Diameter	No Batters Required			
Top	#1	#2	#3	#4
66″	16			
60″	16	16		
54″	16	16	14	
48″	16	16	14	14
42″	16	16	14	26
36″	16	16	14	38
30″	16	16	14	48
24″	16	16	14	58

MAXIMUM REACH OF POWER SHOVELS (FEET)

Size of Dipper (yard)	Length Standard Boom	Length Standard Handle	Maximum Cutting Height	Maximum Cutting Radius
⅜	13 to 15	11 to 13	17 to 19	to 22
½	15 to 16	12 to 13	19 to 24	23 to 24
¾	17 to 18	13 to 15	21 to 27	25 to 28
1	20	16	23 to 27	31 to 32
1¼	21	16	23 to 27	31 to 32
1½	21 to 23	16 to 18	24 to 29	32 to 33
1¾	22 to 24	16 to 18	26 to 30	32 to 33
1	22 to 25	17 to 19	26 to 30	33 to 36
2½	25 to 26	18 to 19	28 to 35	35 to 38

(Standard sizes of power shovels with standard boom and dipper handle lengths, showing maximum cutting length and reach with boom at 45°)

OPTIMUM DEPTH OF CUT (FEET)

Size of Dipper (yard)	Light, free flowing material, such as loam, sand, gravel	Medium materials, such as good common earth	Harder materials such as rough and hard or wet, sticky clay
⅜	3.8	4.5	6.0
½	4.6	5.7	7.0
¾	5.3	6.8	8.0
1	6.0	7.8	9.0
1¼	6.5	8.5	9.8
1½	7.0	9.2	10.7
1¾	7.4	9.7	11.5
2	7.8	10.2	12.2
2½	8.4	11.2	13.3

(Depth of cut to produce the greatest output and at which the dipper comes up with a full load without undue crowding; has nothing to do with maximum digging range of the machine)

EFFECT OF DEPTH OF CUT AND ANGLE OF SWING ON POWER SHOVEL OUTPUT

Depth of Cut in percent of optimum	Angle of Swing in Degrees						
	45°	60°	75°	90°	120°	150°	180°
40%	.93	.89	.85	.80	.72	.65	.59
60%	1.10	1.03	.95	.91	.81	.73	.66
80%	1.22	1.12	1.04	.98	.86	.77	.69
100%	1.26	1.16	1.07	1.00	.88	.79	.71
120%	1.20	1.11	1.03	.97	.86	.77	.70
140%	1.12	1.04	.97	.91	.81	.73	.66
160%	1.03	.96	.90	.85	.75	.67	.62

(Conversion factors which, when applied to output at 90° swing and optimum depth of cut, will give output at other angles of swing or depth of cut)

GENERAL/GENERIC CONVERSIONS

WEIGHTS AND MEASURES—UNITED STATES

LINEAR MEASURE

1 mile = 8 furlongs; 80 chains; 320 rods; 1760 yards; 5280 feet
1 furlong = 10 chains; 220 yards
1 station = 33.3 yards; 100 feet
1 chain = 4 rods; 22 yards; 66 feet; 100 links
1 rod = 5.5 yards; 16.5 feet
1 yard = 3 feet; 36 inches
1 foot = 12 inches

SURVEYOR'S CHAIN MEASURE

1 link = 7.92 inches
1 statute mile = 80 chains

LAND MEASURE

1 township = 36 sections; 36 square miles
1 square mile = 1 section; 640 acres
1 acre = 4840 sq. yards; 43,560 sq. feet; 160 sq. rods
1 square rod = 272-1/4 sq. feet; 30-1/4 sq. yards
1 square yard = 1,296 sq. inches; 9 sq. feet
1 square foot = 144 sq. inches

CUBIC MEASURE

1 cubic yard = 27 cubic feet
1 cord wood = 4×4×8 feet; 128 cu. ft.
1 ton (shipping) = 40 cu. ft.
1 cubic foot = 1728 cubic in.
1 bushel = 2150.42 cubic in.
1 gallon = 231 cu. in.

WEIGHTS COMMERCIAL

1 long ton = 2240 lbs.
1 short ton = 2000 lbs.
1 pound = 16 ounces
1 ounce = 16 drams

TROY WEIGHT FOR GOLD AND SILVER

1 pound = 12 ounces; 5760 grains
1 pennyweight = 24 grains
1 ounce = 20 pennyweights
1 ounce = 480 grains

DRY MEASURE

1 quart = 2 pints; 67.20 cu. in.
1 peck = 16 pints; 537.605 cu. in.
1 bushel = 4 pecks; 32 quarts; 2150.42 cu. in.

MARINER'S MEASURE

1 fathom = 6 feet
1 cable length = 120 fathoms
1 statute mile = 7-1/3 cable lengths; 5,280 feet
1 marine league = 3 marine miles
1 nautical mile = 6,080 feet.

MEASURES OF POWER

1 BTU per minute = .0236 HP; 17.6 watts; .0176 kilowatts; 778 ft. lbs./min.
1 ft. lb. min. = .0226 watts; .001285 BTU per minute
1 horsepower = 746 watts; 33,000 ft. lbs./min.; 42.4 BTU/min.
1 watt = .00134 HP; .001 kilowatts; 44.2 ft. lbs./min.; .0568 BTU/min.
1 kilowatt = 1.341 HP; 1000 watts; 44,250 ft. lbs./min.; 56.8 BTU/min.

LIQUID MEASURE

1 pint = 4 gills; 28.875 cu. in.
1 quart = 2 pints; 57.75 cu. in.
1 hogshead = 63 gallons
1 barrel = 31-1/2 gallons
1 gallon = 4 quarts; 8 pints; 32 gills; 231 cu. in.; 8-1/3 lbs. @ 62 degrees F.
1 cu. ft. water = 7.48 gals.; 1728 cu. in.; 62-1/2 lbs. @ 62 degrees F.

METRIC—U.S. CONVERSION FACTORS (based on National Bureau of Standards)

AREA

Sq. cm. × 0.1550 = sq. ins.
Sq. m. × 10.7639 = sq. ft.
1 are = 100 square meters
Ares × 1076.39 = sq. ft.
Sq. m. × 1.1960 = sq. yds.
Hectare × 2.4710 = acres
Sq. km. × 0.3861 = sq. miles
Sq. ins. × 6.4516 = sq. cm.
Sq. ft. × 0.0929 = sq. m.
Sq. ft. × 0.00093 = ares
Sq. yds. × 0.8361 = sq. m.
Acre × 0.4047 = hectares
Sq. miles × 2.5900 = sq. km.

LENGTH

Centimeters × 0.3937 = inches
Meters × 3.2808 = feet
Meters × 1.0936 = yards
Kilometers × 0.6214 = statute miles
Kilometers × 0.53959 = nautical miles
Inches × 2.5400 = centimeters
Feet × 0.3048 = meters
Yards × 0.9144 = meters
Miles × 1.6093 kilometers
Miles × 1.85325 = kilometers

VOLUME

Cu. cm. × 0.0610 = cu. ins.
Cu. m. × 35.3145 = cu. ft.
Cu. m. × 1.3079 = cu. yds.
Cu. ins. × 16.3872 = cu. cm.
Cu. ft. × 0.0283 = cu. m.
Cu. yds. × 0.7646 = cu. m.

WEIGHT

Grams × 15.4324 = grains
Grams × 0.0353 = oz.
Grams × 0.0022 = lbs.
Kgs. × 2.2046 = lbs.
Kgs. × 0.0011 = tons (short)
Kgs. 0.00098 = tons (long)
Metric Tons × 1.1023 = tons (short)
Metric Tons × 2204.62 = lbs.
Grains × 0.0648 = g.
Oz. × 28.3495 = g.
Lbs. × 453.592 = g.
Lbs. × 0.4536 = kg.
Lbs. × 0.0004536 = tons metric
Tons (short) × 907.1848 = kg.
Tons (short) × 0.9072 = metric tons
Tons (long) × 1016.05 = kg.

CAPACITY

Liters × 61.0250 = cu. in.
Liters × 0.0353 = cu. ft.
Liters × 0.2642 = gals. (U.S.)
Liters × 0.0284 = bushels (U.S.)
Liters × 1000.027 = cu. cm.
Liters × 1.0567 = qt. (liquid) or 0.9081 = qt. (dry)
Liters × 2.2046 = lb. of pure water at 4° C = 1 kg.
Cu. ins. × 0.0164 = liters
Cu. ft. × 28.3162 = liters
Gallons × 3.7853 = liters
Bushels × 35.2383 = liters

PRESSURE

Kgs. per sq. cm. × 14.223 = lbs. per sq. in.
Lbs. per sq. in. × 0.0703 = kgs. per sq. cm.
Kgs. per sq. in × 0.2048 = lbs. per sq. ft.
Kgs. per sq. m. × 0.204817 = lbs. per sq. ft.
Lbs. per sq. ft. × 4.8824 = kgs. per sq. m.
Kgs. per sq. m. × 0.00009144 = tons (long) per sq. ft.
Tons long per sq. ft. × 0.0001094 = kg. per sq. m.
Kgs. per sq. mm. × 0.634973 = tons (long) per sq. in.
Tons long per sq. ft. × 0.0001094 = kg. per sq. m.
Kgs. per cu. m. × 0.062428 = lbs. per cu. ft.
Lbs. per cu. ft. × 16.0184 = kgs. per cu. m.
Kgs. per m. × 0.671972 = lbs. per ft.
Lbs. per ft. × 1.48816 = kgs. per m.
Kg.-m. × 7.233 = ft.-lbs.
Ft.-lbs. × 0.13826 = kg.-m.

POWER

Metric horsepower × .98632 = U.S. horsepower
U.S. horsepower × 1.01387 = metric horsepower

CONVERSION TABLE

Inches		Centimeters	Centimeters		Inches
1	=	2.54001	1	=	0.39370
2	=	5.08001	2	=	0.78740
3	=	7.62002	3	=	1.1811
4	=	10.16002	4	=	1.5748
5	=	12.70003	5	=	1.9685
6	=	15.24003	6	=	2.3622
7	=	17.78004	7	=	2.7559
8	=	20.32004	8	=	3.1496
9	=	22.86005	9	=	3.5433
Feet		**Meters**	**Meters**		**Feet**
1	=	0.304801	1	=	3.28083
2	=	0.609601	2	=	6.56167
3	=	0.914402	3	=	9.84250
4	=	1.219202	4	=	13.12333
5	=	1.524003	5	=	16.40417
6	=	1.828804	6	=	19.68500
7	=	2.133604	7	=	22.96583
8	=	2.438405	8	=	26.24666
9	=	2.743205	9	=	29.52750
Yards		**Meters**	**Meters**		**Yards**
1	=	0.914402	1	=	1.093611
2	=	1.828804	2	=	2.187222
3	=	2.743205	3	=	3.280833
4	=	3.657607	4	=	4.374444
5	=	4.572009	5	=	5.468056
6	=	5.486411	6	=	6.561667
7	=	6.400813	7	=	7.655278
8	=	7.315215	8	=	8.748889
9	=	8.229616	9	=	9.842500
Miles		**Kilometers**	**Kilometers**		**Miles**
1	=	1.60935	1	=	0.62137
2	=	3.21869	2	=	1.24274
3	=	4.82804	3	=	1.86411
4	=	6.43739	4	=	2.48548
5	=	8.04674	5	=	3.10685
6	=	9.65608	6	=	3.72822
7	=	11.26543	7	=	4.34959
8	=	12.87478	8	=	4.97096
9	=	14.48412	9	=	5.59233
Pounds Av.		**Kilograms**	**Kilograms**		**Pounds Av.**
1	=	0.45359	1	=	2.20462
2	=	0.90718	2	=	4.40924
3	=	1.36078	3	=	6.61387
4	=	1.81437	4	=	8.81849
5	=	2.26796	5	=	11.02311
6	=	2.72155	6	=	13.22773
7	=	3.17514	7	=	15.43236
8	=	3.62874	8	=	17.63698
9	=	4.08233	9	=	19.84160

DECIMALS OF AN INCH

Fraction	64ths	Decimal	Fraction	64ths	Decimal
			1/2	32	.500
—	1	.015625	—	33	.515625
1/32	2	.03125	17/32	34	.53125
—	3	.046875	—	35	.546875
1/16	4	.0625	9/16	36	.5625
—	5	.078125	—	37	.578125
3/32	6	.09375	19/32	38	.59375
—	7	.109375	—	39	.609375
1/8	8	.125	5/8	40	.625
—	9	.140625	—	41	.640625
5/32	10	.15625	21/32	42	.65625
—	11	.171875	—	43	.671875
3/16	12	.1875	11/16	44	.6875
—	13	.203125	—	45	.703125
7/32	14	.21875	22/32	46	.71875
—	15	.234375	—	47	.734375
1/4	16	.250	3/4	48	.750
—	17	.265625	—	49	.765625
9/32	18	.28125	25/32	50	.78125
—	19	.296875	—	51	.796875
5/16	20	.3125	13/16	52	.8125
—	21	.328125	—	53	.828125
11/32	22	.34375	27/32	54	.84375
—	23	.359375	—	55	.859375
3/8	24	.375	7/8	56	.875
—	25	.390625	—	57	.890625
13/32	26	.40625	29/32	58	.90625
—	27	.421875	—	59	.921875
7/16	28	.4375	15/16	60	.9375
—	29	.453125	—	61	.953125
15/32	30	.46875	31/32	62	.96875
—	31	.484375	—	63	.984375

DECIMALS OF A FOOT

0″	.0000	**1″**	.0833	**2″**	.166667	**3″**	.2500
1/16	.0052	1 1/16	.0885	2 1/16	.171875	3 1/16	.2552
1/8	.0104	1 1/8	.09375	2 1/8	.1771	3 1/8	.2604
3/16	.015625	1 3/16	.0990	2 3/16	.1823	3 3/16	.265625
1/4	.0208	1 1/4	.1042	2 1/4	.1875	3 1/4	.2708
5/16	.0260	1 5/16	.109375	2 5/16	.1927	3 5/16	.2760
3/8	.03125	1 3/8	.1146	2 3/8	.1979	3 3/8	.28125
7/16	.0365	1 7/16	.1198	2 7/16	.203125	3 7/16	.2865
1/2	.0417	1 1/2	.1250	2 1/2	.2083	3 1/2	.2917
9/16	.046875	1 9/16	.1302	2 9/16	.2135	3 9/16	.296875
5/8	.0521	1 5/8	.1354	2 5/8	.21875	3 5/8	.3021
11/16	.0573	1 11/16	.140625	2 11/16	.2240	3 11/16	.3073
3/4	.0625	1 3/4	.1458	2 3/4	.2292	3 3/4	.3125
13/16	.0677	1 13/16	.1510	2 13/16	.234375	3 13/16	.3177
7/8	.0729	1 7/8	.15625	2 7/8	.2396	3 7/8	.3229
15/16	.078125	1 15/16	.1615	2 15/16	.2448	3 15/16	.328125
4″	.3333	**5″**	.416667	**6″**	.5000	**7″**	.5833
4 1/16	.3385	5 1/16	.421875	6 1/16	.5052	7 1/16	.5885
4 1/8	.34375	5 1/8	.4271	6 1/8	.5104	7 1/8	.59375
4 3/16	.3490	5 3/16	.4323	6 3/16	.515625	7 3/16	.5990
4 1/4	.3542	5 1/4	.4375	6 1/4	.5208	7 1/4	.6042
4 5/16	.359375	5 5/16	.4427	6 5/16	.5260	7 5/16	.6093
4 3/8	.3646	5 3/8	.4479	6 3/8	.53125	7 3/8	.6146
4 7/16	.3698	5 7/16	.453125	6 7/16	.5365	7 7/16	.6198
4 1/2	.3750	5 1/2	.4583	6 1/2	.5417	7 1/2	.6250
4 9/16	.3802	5 9/16	.4635	6 9/16	.546875	7 9/16	.6302
4 5/8	.3854	5 5/8	.46875	6 5/8	.5521	7 5/8	.6354
4 11/16	.390625	5 11/16	.4740	6 11/16	.5573	7 11/16	.640625
4 3/4	.3958	5 3/4	.4792	6 3/4	.5625	7 3/4	.6458
4 13/16	.4010	5 13/16	.484375	6 13/16	.5677	7 13/16	.6510
4 7/8	.40625	5 7/8	.4896	6 7/8	.5729	7 7/8	.65625
4 15/16	.4115	5 15/16	.4948	6 15/16	.578125	7 15/16	.6615
8″	.666667	**9″**	.7500	**10″**	.8333	**11″**	.916667
8 1/16	.671875	9 1/16	.7552	10 1/16	.8385	11 1/16	.921875
8 1/8	.6771	9 1/8	.7604	10 1/8	.84375	11 1/8	.9271
8 3/16	.6823	9 3/16	.765625	10 3/16	.8490	11 3/16	.9323
8 1/4	.6875	9 1/4	.7708	10 1/4	.8542	11 1/4	.9375
8 5/16	.6927	9 5/16	.7760	10 5/16	.859375	11 5/16	.9427
8 3/8	.6979	9 3/8	.78125	10 3/8	.8646	11 3/8	.9479
8 7/16	.703125	9 7/16	.7865	10 7/16	.8698	11 7/16	.953125
8 1/2	.7083	9 1/2	.7917	10 1/2	.8750	11 1/2	.9583
8 9/16	.7135	9 9/16	.796875	10 9/16	.8802	11 9/16	.9635
8 5/8	.71875	9 5/8	.8021	10 5/8	.8854	11 5/8	.96875
8 11/16	.7240	9 11/16	.8073	10 11/16	.890625	11 11/16	.9740
8 3/4	.7292	9 3/4	.8125	10 3/4	.8958	11 3/4	.9792
8 13/16	.734375	9 13/16	.8177	10 13/16	.9010	11 13/16	.984375
8 7/8	.7396	9 7/8	.8229	10 7/8	.90625	11 7/8	.9896
8 15/16	.7448	9 15/16	.828125	10 15/16	.9115	11 15/16	.9948

SIMPLE AND COMPOUND INTEREST RATE TABLES

These tables give you the amount accumulated after a specified period at various simple interest rates.

How to Use These Tables

1. First find the column headed by the interest rate on the investment in question.
2. Then scan down to the appropriate number of years. At that intersection, you will find a factor.
3. Multiply that factor by the principal. The result is the amount accumulated within that period.

Example: Lester Mannington has $600 in a savings account at Federal Bank and Trust Company that is earning 7% simple interest per year. He wants to know how much he will have in the bank at the end of 10 years.

1. Across from 10 years and under the 7% column, you will find the factor 1.70.
2. Multiply this factor (1.70) by the $600 to get $1,020, the value of the account after 10 years at simple interest.

SIMPLE INTEREST RATE

Number of Years	7%	8%	9%	10%	11%	12%	13%	14%	15%	20%
1	1.07	1.08	1.09	1.10	1.11	1.12	1.13	1.14	1.15	1.20
2	1.14	1.16	1.18	1.20	1.22	1.24	1.26	1.28	1.30	1.40
3	1.21	1.24	1.27	1.30	1.33	1.36	1.39	1.42	1.45	1.60
4	1.28	1.32	1.36	1.40	1.44	1.48	1.52	1.56	1.60	1.80
5	1.35	1.40	1.45	1.50	1.55	1.60	1.65	1.70	1.75	2.00
6	1.42	1.48	1.54	1.60	1.66	1.72	1.78	1.84	1.90	2.20
7	1.49	1.56	1.63	1.70	1.77	1.84	1.91	1.98	2.05	2.40
8	1.56	1.64	1.72	1.80	1.88	1.96	2.04	2.12	2.20	2.60
9	1.63	1.72	1.81	1.90	1.99	2.08	2.17	2.26	2.35	2.80
10	1.70	1.80	1.90	2.00	2.10	2.20	2.30	2.40	2.50	3.00
11	1.77	1.88	1.99	2.10	2.21	2.32	2.43	2.54	2.65	3.20
12	1.84	1.96	2.08	2.20	2.32	2.44	2.56	2.68	2.80	3.40
13	1.91	2.04	2.17	2.30	2.43	2.56	2.69	2.82	2.95	3.60
14	1.98	2.12	2.26	2.40	2.54	2.68	2.82	2.96	3.10	3.80
15	2.05	2.20	2.35	2.50	2.65	2.80	2.95	3.10	3.25	4.00
16	2.12	2.28	2.44	2.60	2.76	2.92	3.08	3.24	3.40	4.20
17	2.19	2.36	2.53	2.70	2.87	3.04	3.21	3.38	3.55	4.40
18	2.26	2.44	2.62	2.80	2.98	3.16	3.34	3.52	3.70	4.60
19	2.33	2.52	2.71	2.90	3.09	3.28	3.47	3.66	3.85	4.80
20	2.40	2.60	2.80	3.00	3.20	3.40	3.60	3.80	4.00	5.00
21	2.47	2.68	2.89	3.10	3.31	3.52	3.73	3.94	4.15	5.20
22	2.54	2.76	2.98	3.20	3.42	3.64	3.86	4.08	4.30	5.40
23	2.61	2.84	3.07	3.30	3.53	3.76	3.99	4.22	4.45	5.60
24	2.68	2.92	3.16	3.40	3.64	3.88	4.12	4.36	4.60	5.80
25	2.75	3.00	3.25	3.50	3.75	4.00	4.25	4.50	4.75	6.00
26	2.82	3.08	3.34	3.60	3.86	4.12	4.38	4.64	4.90	6.20
27	2.89	3.16	3.43	3.70	3.97	4.24	4.51	4.78	5.05	6.40
28	2.96	3.24	3.52	3.80	4.08	4.36	4.64	4.92	5.20	6.60
29	3.03	3.32	3.61	3.90	4.19	4.48	4.77	5.06	5.35	6.80
30	3.10	3.40	3.70	4.00	4.30	4.60	4.90	5.20	5.50	7.00
31	3.17	3.48	3.79	4.10	4.41	4.72	5.03	5.34	5.65	7.20
32	3.24	3.56	3.88	4.20	4.52	4.84	5.16	5.48	5.80	7.40
33	3.31	3.64	3.97	4.30	4.63	4.96	5.29	5.62	5.95	7.60
34	3.38	3.72	4.06	4.40	4.74	5.08	5.42	5.76	6.10	7.80
35	3.45	3.80	4.15	4.50	4.85	5.20	5.55	5.90	6.25	8.00
36	3.52	3.88	4.24	4.60	4.96	5.32	5.68	6.04	6.40	8.20
37	3.59	3.96	4.33	4.70	5.07	5.44	5.81	6.18	6.55	8.40
38	3.66	4.04	4.42	4.80	5.18	5.56	5.94	6.32	6.70	8.60
39	3.73	4.12	4.51	4.90	5.29	5.68	6.07	6.46	6.85	8.80
40	3.80	4.20	4.60	5.00	5.40	5.80	6.20	6.60	7.00	9.00

COMPOUND INTEREST RATE

Number of Years	7%	8%	9%	10%	11%	12%	13%	14%	15%	20%
1	1.0700	1.0800	1.0900	1.1000	1.1100	1.1200	1.1300	1.1400	1.1500	1.2000
2	1.1449	1.1664	1.1881	1.2100	1.2321	1.2544	1.2769	1.2996	1.3225	1.4400
3	1.2250	1.2597	1.2950	1.3310	1.3576	1.4049	1.4428	1.4815	1.5208	1.7280
4	1.3107	1.3604	1.4115	1.4647	1.5180	1.5735	1.6304	1.6389	1.7409	2.0736
5	1.4025	1.4693	1.5386	1.6105	1.6350	1.7623	1.8424	1.9254	2.0113	2.4883
6	1.5007	1.5868	1.6771	1.7715	1.8704	1.9738	2.0819	2.1949	2.3130	2.9859
7	1.6057	1.7138	1.8230	1.9487	2.0761	2.2106	2.3526	2.5022	2.6600	3.5831
8	1.7181	1.8509	1.9925	2.1435	2.3045	2.4759	2.6584	2.8525	3.0590	4.2998
9	1.8384	1.9990	2.1718	2.3579	2.5580	2.7730	3.0040	3.2519	3.5178	5.1597
10	1.9671	2.1589	2.3673	2.5937	2.8394	3.1058	3.3945	3.7072	4.0455	6.1917
11	2.1048	2.3316	2.5804	2.8531	3.1517	3.4785	3.8358	4.2262	4.6523	7.4300
12	2.2521	2.5181	2.8126	3.1384	3.4984	3.8959	4.3345	4.8179	5.3502	8.9161
13	2.4098	2.7196	3.0658	3.4522	3.8832	4.3634	4.8980	5.4924	6.1527	10.6993
14	2.5785	2.9371	3.3417	3.7974	4.3104	4.8871	5.5347	6.2613	7.0757	12.8391
15	2.7590	3.1721	3.6424	4.1772	4.7845	5.4735	6.2542	7.1379	8.1370	15.4270
16	2.9521	3.4259	3.9703	4.5949	5.3108	6.1303	7.0673	8.1372	9.3576	18.4884
17	3.1588	3.7000	4.3276	5.0544	5.8950	6.8660	7.9860	9.2764	10.6712	22.1861
18	3.3799	3.9960	4.7171	5.5599	6.5435	7.6899	9.0242	10.5751	12.3754	26.6233
19	3.6165	4.3157	5.1416	6.1159	7.2633	8.6127	10.1974	12.0556	14.2317	31.9479
20	3.8696	4.6609	5.6044	6.7274	8.0623	9.6462	11.5230	13.7434	16.3665	38.3375
21	4.1405	5.0338	6.1088	7.4002	8.9491	10.8038	13.0210	15.6675	18.8215	46.0051
22	4.4304	5.4365	6.6586	8.1402	9.9335	12.1003	14.7138	17.8610	21.6447	55.2061
23	4.7405	5.8714	7.2578	8.9543	11.0262	13.5523	16.6266	20.3615	24.8914	66.2473
24	5.0723	6.3411	7.9110	9.8497	12.2391	15.1786	18.7880	23.2122	28.6251	79.4968
25	5.4274	6.8484	8.6230	10.8347	13.5854	17.0000	21.2305	26.4619	32.9189	95.3962
26	5.8073	7.3963	9.3991	11.9181	15.0793	19.0400	23.9905	30.1665	37.8567	114.4754
27	6.2138	7.9880	10.2450	13.1099	16.7386	21.3248	27.1092	34.3899	43.5353	137.3705
28	6.6488	8.6271	11.1671	14.4209	18.5799	23.8838	30.6334	39.2044	50.0656	164.8446
29	7.1142	9.3172	12.1271	15.8630	20.6236	26.7499	34.6158	44.6931	57.5754	197.8135
30	7.6122	10.5582	13.2676	17.4494	22.8922	29.9599	39.1158	50.9501	66.2117	237.3763
31	8.1451	10.8676	14.4617	19.1943	25.4104	33.5551	44.2009	58.0831	76.1435	284.8515
32	8.7152	11.7370	15.7633	21.1137	28.2055	37.5817	49.9470	66.2148	87.5650	341.8218
33	9.3253	12.6760	17.1820	23.2251	31.3082	42.0915	56.4402	75.4849	100.6998	410.1862
34	9.9781	13.6901	18.7284	25.5476	34.7521	47.1425	63.7774	86.0527	115.8048	492.2235
35	10.6765	14.7853	20.4139	28.1024	38.5748	52.7996	72.0685	98.1001	133.1755	590.6682
36	11.4239	15.9681	22.2512	30.9128	42.8180	59.1355	81.4374	111.8342	153.1518	708.8018
37	12.2236	17.2456	24.2538	34.0039	47.5280	66.2318	92.0242	127.4909	176.1246	850.5622
38	13.0792	18.6252	26.4366	37.4048	52.7561	74.1796	103.9874	145.3397	202.5433	1020.6746
39	13.9948	20.1152	28.8159	41.1447	58.5593	83.0812	117.5057	165.6872	232.9248	1224.8096
40	14.9744	21.7245	31.4094	45.2592	65.0008	93.0509	132.7815	188.8835	267.8635	1469.7715

ROAD MILEAGE CHART BETWEEN U.S. CITIES

	Atlanta	Baltimore	Birmingham	Boston	Buffalo	Chicago	Cincinnati	Cleveland	Dallas	Denver	Des Moines	Detroit	Houston	Indianapolis	Kansas City, Mo.	Los Angeles	Memphis	Milwaukee	Minneapolis	Nashville	New Orleans	New York City	Philadelphia	Phoenix	Pittsburgh	St. Louis	Salt Lake City	San Francisco	Seattle	Washington, D.C.
Atlanta		669	152	1071	877	695	461	686	805	1401	894	726	814	508	810	2197	366	784	1105	256	493	855	766	1810	697	558	1900	2523	2756	630
Baltimore	669		787	399	345	687	494	351	1458	1701	1025	511	1449	580	1099	2695	947	776	1097	725	1138	187	97	2325	230	817	2118	2876	2748	39
Birmingham	152	787		1185	902	636	476	716	653	1282	808	741	662	480	706	2056	247	745	1066	205	351	974	884	1658	742	476	1781	2393	2575	748
Boston	1068	399	1185		449	975	876	632	1819	1989	1311	*699	1916	941	1456	3052	1355	1064	1385	1165	1536	216	304	2682	598	1178	2405	3163	3036	437
Chicago	695	687	656	975	529		295	343	936	1016	338	275	1085	187	499	2095	544	91	412	451	929	843	762	1722	461	291	1431	2189	2063	687
Cincinnati	461	494	476	876	430	295		244	943	1169	571	265	1040	108	590	2186	479	384	705	290	820	659	578	1816	284	338	1644	2402	2356	492
Cleveland	686	351	716	632	186	343	244		1187	1357	679	167	1284	309	824	2420	723	432	753	533	1063	507	426	2050	125	546	1772	2530	2404	351
Dallas	805	1458	653	1819	1373	941	943	1187		784	704	1188	242	882	499	1403	464	1015	956	686	498	1607	1526	1005	1232	645	1241	1806	2112	1372
Denver	1401	1701	1282	1989	1543	1016	1169	1357	784		679	1284	1026	1061	604	1134	1035	1040	841	1158	1282	1851	1770	818	1482	856	512	1270	1347	1696
Detroit	726	511	741	*699	*252	275	265	167	1188	1284	606		1337	277	751	2347	713	364	685	555	1085	667	586	1977	285	543	1700	2458	2336	511
Houston	814	1449	662	1916	1470	1085	1040	1284	242	1026	946	1337		997	741	1553	561	1174	1198	783	359	1636	1546	1153	1319	794	1431	1955	2302	1410
Indianapolis	508	580	480	941	495	187	108	309	882	1061	483	277	997		484	2080	436	276	597	294	805	729	648	1710	354	237	1536	2294	2245	564
Jacksonville	314	793	426	1191	1067	1009	775	952	994	1708	1208	996	911	822	1123	2397	673	1098	1419	570	572	980	890	1999	865	872	2207	2799	3070	754
Los Angeles	2197	2695	2056	3052	2606	2095	2186	2420	1403	1134	1801	2347	1553	2080	1596		1831	2176	1940	2058	1901	2915	2721	398	2533	1848	734	403	1145	2644
Memphis	366	947	247	1355	909	544	479	723	464	1035	608	713	561	436	459	1831		624	840	222	401	1138	1057	1466	758	294	1534	2157	2331	908
Mexico City	1800	2435	1648	2902	2456	2071	2026	2270	1149	1754	1852	2337	986	1983	1647	2017	1547	2164	2104	1769	1345	2622	2532	1625	2305	1794	2100	2419	2948	2396
Miami	665	1144	765	1542	1418	1360	1126	1303	1309	2046	1559	1347	1216	1173	1470	2712	1011	1449	1770	921	875	1330	1241	2314	1216	1223	2545	3075	3421	1105
Minneapolis	1105	1097	1066	1385	939	412	705	753	956	841	252	685	1198	597	457	1940	840	326		861	1241	1253	1172	1630	871	546	1239	1997	1641	1097
Nashville	256	725	205	1165	719	451	290	533	686	1158	638	555	783	294	554	2058	222	540	861		530	949	868	1725	559	302	1670	2410	2512	686
New Orleans	493	1138	351	1536	1253	925	820	1063	498	1282	989	1085	359	805	821	1901	401	1013	1241	530		1325	1235	1503	1093	695	1739	2303	2610	1099
New York City	855	187	974	216	445	843	658	507	1607	1851	1173	667	1636	729	1319	2915	1138	932	1253	949	1325		92	2459	386	966	2267	3025	2904	225
Omaha	1012	1163	911	1449	1003	476	691	817	656	540	139	744	898	583	205	1662	664	500	358	756	1026	1312	1230	1335	942	454	955	1713	1667	1156
Philadelphia	766	97	884	304	365	762	578	426	1526	1770	1092	586	1546	648	1238	2721	1057	851	1172	868	1235	92		2464	305	885	2186	2944	2823	136
Phoenix	1810	2325	1658	2682	2236	1722	1816	2050	1005	818	1430	1977	1155	1710	1226	398	1466	1806	1630	1725	1503	2459	2464		2163	1478	653	800	1541	2274
Pittsburgh	697	230	742	598	217	461	284	125	1232	1482	804	285	1319	354	937	2533	758	550	871	559	1093	386	305	2163		591	1890	2648	2522	230
St. Louis	558	817	476	1178	802	291	338	546	645	856	336	543	794	237	252	1848	294	370	546	302	695	966	885	1478	591		1368	2126	2109	801
Salt Lake City	1900	2118	1781	2405	1958	1431	1644	1772	1241	512	1094	1700	1431	1536	1116	734	1534	1455	1239	1670	1739	2267	2186	653	1890	1368		759	871	2111
San Francisco	2523	2876	2393	3163	2716	2189	2402	2530	1806	1270	1852	2458	1955	2294	1874	403	2157	2213	1997	2410	2303	3025	2944	800	2648	2126	759		827	2869
Seattle	2756	2748	2575	3036	2590	2063	2356	2404	2112	1347	1773	2336	2302	2248	1872	1145	2331	1977	1641	2512	2610	2904	2823	1541	2522	2109	871	827		2748
Washington, D.C	630	39	748	417	359	687	492	351	1372	1696	1018	511	1410	564	1048	2644	908	776	1097	686	1099	225	136	2274	230	801	2111	2869	2748	

U.S. TELEPHONE AREA CODES

Area Codes by the Numbers *(Includes new area codes for 2000)*

Code	State	Principal Areas
201	New Jersey	Hackensack, Jersey City
202	DC	Washington, D.C.
203	Connecticut	New Haven, Stamford, southwestern region
205	Alabama	Birmingham
206	Washington	Seattle
207	Maine	Entire state
208	Idaho	Entire state
209	California	Modesto, Stockton
210	Texas	San Antonio
212	New York	New York City/Manhattan
213	California	Los Angeles (downtown business district)
214	Texas	Dallas (central zone)
215	Pennsylvania	Philadelphia
216	Ohio	Cleveland
217	Illinois	Champaign, Springfield
218	Minnesota	Duluth, northern region
219	Indiana	Ft. Wayne, northern region
224	Illinois	Northern Chicago (overlays 847)
225	Louisiana	Baton Rouge
228	Mississippi	Biloxi, Gulfport, southern region
231	Michigan	Muskegon, Traverse City (overlays 616)
240	Maryland	Silver Spring, Frederick (overlays 301)
248	Michigan	Pontiac, Troy, northwestern Detroit suburbs
252	North Carolina	Greenville, Rocky Mount, Cape Hatteras
253	Washington	Tacoma, southern Seattle area
254	Texas	Waco
256	Alabama	Huntsville, northern and east central regions
262	Wisconsin	Kenosha, Racine, West Bend (split from 414)
267	Pennsylvania	Philadelphia (overlays 215)
270	Kentucky	Bowling Green (split from 502)
281	Texas	Houston outer suburbs
301	Maryland	Silver Spring, Washington, D.C. suburbs, Frederick
302	Delaware	Entire state
303	Colorado	Boulder, Denver (overlays 720)
304	West Virginia	Entire state
305	Florida	Miami, Key West
307	Wyoming	Entire state
308	Nebraska	North Platte, Grand Island, western region

U.S. TELEPHONE AREA CODES

Code	State	Principal Areas
309	Illinois	Bloomington, Peoria, Moline/Quad Cities
310	California	Santa Monica, LAX Airport
312	Illinois	Chicago (downtown)
313	Michigan	Detroit
314	Missouri	St. Louis
315	New York	Syracuse, Utica, north central region
316	Kansas	Wichita, southern half region
317	Indiana	Indianapolis
318	Louisiana	Shreveport, Lafayette, western region
319	Iowa	Cedar Rapids, Davenport/Quad Cities, eastern region
320	Minnesota	St. Cloud, central and west central regions
321	Florida	Orlando (overlays 407)
323	California	Los Angeles (outer downtown)
330	Ohio	Akron, Canton, Youngstown
334	Alabama	Montgomery, Mobile, southern region
336	North Carolina	Greensboro, High Point, Winston-Salem
337	Louisiana	Lafayette, Lake Charles (split from 318)
341	California	Oakland (overlays 510)
347	New York	Bronx, Brooklyn, Queens, Staten Island (overlays 718)
352	Florida	Gainesville, Ocala
360	Washington	Bellingham, Olympia, Vancouver
401	Rhode Island	Entire state
402	Nebraska	Omaha, eastern region
404	Georgia	Atlanta
405	Oklahoma	Oklahoma City, central region
406	Montana	Entire state
407	Florida	Orlando
408	California	San Jose
409	Texas	Galveston, Beaumont, suburbs around Houston, southeastern region
410	Maryland	Baltimore, northern and eastern regions
412	Pennsylvania	Pittsburgh
413	Massachusetts	Springfield, western region
414	Wisconsin	Milwaukee
415	California	San Francisco, Marin County
417	Missouri	Springfield, Joplin, southwestern region
419	Ohio	Toledo, northwestern region
423	Tennessee	Knoxville, Chattanooga, eastern region
424	California	Santa Monica, LAX Airport (overlays 310)
425	Washington	Everett, Redmond, northern and eastern Seattle area
435	Utah	Logan, entire state except Salt Lake City area

U.S. TELEPHONE AREA CODES

Code	State	Principal Areas
440	Ohio	Lorain, Ashtabula, suburbs near Cleveland
443	Maryland	Baltimore, northern and eastern region (overlays 410)
469	Texas	Dallas and suburbs (overlays 214 and 972)
484	Pennsylvania	Allentown, King of Prussia (overlays 610)
500	—	Personal Communications Services
501	Arkansas	Little Rock, Ft. Smith, central and northwestern regions
502	Kentucky	Louisville, western region
503	Oregon	Portland, Salem, Astoria, northwestern region
504	Louisiana	New Orleans, Hammond
505	New Mexico	Entire state
507	Minnesota	Rochester, southern region
508	Massachusetts	Worcester, Cape Cod, southeastern region
509	Washington	Spokane, eastern region
510	California	Oakland
512	Texas	Austin, Corpus Christi
513	Ohio	Cincinnati
515	Iowa	Des Moines, central region
516	New York	Long Island
517	Michigan	Lansing, Saginaw
518	New York	Albany, northeastern region
520	Arizona	Tucson, Flagstaff, Yuma, all of state except Phoenix area
530	California	Redding, Lake Tahoe, northeastern region
533	—	Personal Communications Services
540	Virginia	Roanoke, western region
541	Oregon	Eugene, all of state except Portland, Salem area
559	California	Fresno
561	Florida	West Palm Beach, Boca Raton
562	California	Long Beach
571	Virginia	Arlington, Washington, D.C. suburbs (overlays 703)
573	Missouri	Columbia, central and western region except St. Louis
580	Oklahoma	Lawton, panhandle, southern, western, north central regions
601	Mississippi	Jackson, all of state except extreme southern tip
602	Arizona	Phoenix metro area
603	New Hampshire	Entire state
605	South Dakota	Entire state
606	Kentucky	Lexington, Covington, eastern region
607	New York	Binghamton, south central region
608	Wisconsin	Madison, southwestern region
609	New Jersey	Atlantic City, Trenton, Camden, southern region
610	Pennsylvania	Allentown, King of Prussia

U.S. TELEPHONE AREA CODES

Code	State	Principal Areas
612	Minnesota	Minneapolis
614	Ohio	Columbus
615	Tennessee	Nashville
616	Michigan	Grand Rapids, Kalamazoo
617	Massachusetts	Boston
618	Illinois	East St. Louis, southern region
619	California	San Diego
626	California	Pasadena, San Gabriel Valley, part of LA County
628	California	San Francisco, Marin County (overlays 415)
630	Illinois	Hinsdale, some western suburbs of Chicago
631	New York	Suffolk County (split of 516)
646	New York	New York City/Manhattan (overlays 212)
650	California	San Mateo, peninsula, San Francisco Airport
651	Minnesota	St. Paul, eastern Twin Cities area
657	California	Anaheim, northern Orange County (overlays 714)
660	Missouri	Sedalia, north central region
661	California	Bakersfield, northern LA County
662	Mississippi	Northern half (split of 516)
669	California	San Jose (overlays 408)
678	Georgia	Atlanta area (overlays 404 and 770)
700	—	Special purposes
701	North Dakota	Entire state
702	Nevada	Las Vegas, Clark County
703	Virginia	Arlington, suburbs of Washington, D.C.
704	North Carolina	Charlotte
706	Georgia	Athens, Columbus, Augusta, Rome
707	California	Santa Rosa, Napa, Eureka, north coastal region
708	Illinois	La Grange, some southern/western Chicago suburbs
710	—	Federal Government uses
712	Iowa	Sioux City, Council Bluffs, western region
713	Texas	Houston/central area
714	California	Anaheim, northern Orange County
715	Wisconsin	Eau Claire, northern/northwestern region
716	New York	Buffalo, Rochester, western region
717	Pennsylvania	Harrisburg
718	New York	Bronx, Brooklyn, Queens, Staten Island
719	Colorado	Colorado Springs, southeastern region
720	Colorado	Boulder, Denver (overlays 303)
724	Pennsylvania	New Castle, western, surrounds Pittsburgh
727	Florida	Proposed overlay to 813

U.S. TELEPHONE AREA CODES

Code	State	Principal Areas
732	New Jersey	New Brunswick, northern Jersey Shore region
734	Michigan	Ann Arbor, Monroe, southern Detroit suburbs, Metro Airport
740	Ohio	Marion, Zanesville, southeastern region
752	California	San Bernardino, Ontario (overlays 909)
757	Virginia	Norfolk
760	California	Escondido, Palm Springs, Bishop, southeastern region
764	California	San Mateo, SF Airport (overlays 909)
765	Indiana	Lafayette, Muncie
770	Georgia	Marietta, Atlanta suburbs
773	Illinois	Chicago outside downtown
775	Nevada	Reno, Carson City, Pahrump, all except Clark County and Las Vegas
781	Massachusetts	Lexington, Boston suburbs
785	Kansas	Topeka, north central, northwestern region
786	Florida	Proposed overlay to 305
800	—	Toll-free services
801	Utah	Salt Lake City, Provo, Ogden, Wasatch Front region
802	Vermont	Entire state
803	South Carolina	Columbia, central region
804	Virginia	Richmond
805	California	San Luis Obispo, Oxnard
806	Texas	Amarillo, Lubbock, panhandle region
808	Hawaii	Entire state
810	Michigan	Flint, Port Huron, northeastern Detroit suburbs
812	Indiana	Evansville, southern region
813	Florida	Tampa, St. Petersburg
814	Pennsylvania	Altoona, Erie
815	Illinois	Joliet, northern region
816	Missouri	Kansas City, St. Joseph
817	Texas	Ft. Worth
818	California	San Fernando
828	North Carolina	Asheville, western tip of state
830	Texas	Del Rio, surrounding San Antonio
831	California	Monterey, Santa Cruz
843	South Carolina	Charleston, coastal region
847	Illinois	Skokie, northern suburbs of Chicago
850	Florida	Tallahassee, panhandle region
856	New Jersey	Camden, Vineland (split of 609)
858	California	San Diego (split of 619)
860	Connecticut	Hartford, northern and eastern region
864	South Carolina	Greenville, Spartanburg, northwestern region

U.S. TELEPHONE AREA CODES

Code	State	Principal Areas
865	Tennessee	Knoxville metro (split of 423)
870	Arkansas	Jonesboro, Pine Bluff, southern, eastern, north central state
877	—	Toll-free services
888	—	Toll-free services
900	—	Premium pay-per-call services
901	Tennessee	Memphis, western region
903	Texas	Tyler, Texarkana, northeastern region
904	Florida	Jacksonville, northeastern region
906	Michigan	Sault Ste. Marie, Upper Peninsula
907	Alaska	Entire state
908	New Jersey	Elizabeth
909	California	San Bernardino, Riverside
910	North Carolina	Wilmington, southeastern region
912	Georgia	Savannah, Macon, southern region
913	Kansas	Kansas City, Leavenworth
914	New York	White Plains
915	Texas	El Paso, Abilene, western region
916	California	Sacramento
917	New York	NYC cellular/pagers/voicemail (overlays 212 and 718)
918	Oklahoma	Tulsa, northeastern region
919	North Carolina	Chapel Hill, Durham, Raleigh
920	Wisconsin	Green Bay, northeastern region
925	California	Concord, Walnut Creek, inland East Bay
931	Tennessee	Clarksville, central except Nashville
935	California	Coronado, southern/eastern suburbs of San Diego (split of 619)
937	Ohio	Dayton
940	Texas	Wichita Falls, Denton
941	Florida	Sarasota, southwestern region
949	California	Irvine, Newport Beach, southern Orange County
951	California	Riverside (split of 909)
952	Minnesota	Minneapolis area (split of 612)
954	Florida	Ft. Lauderdale, Broward County
956	Texas	Brownsville, Laredo, Rio Grande Valley
970	Colorado	Grand Junction, northern and western region
971	Oregon	Portland, Salem (overlays 503)
972	Texas	Dallas suburbs, portions of Dallas
973	New Jersey	Newark, Paterson, Morristown
978	Massachusetts	Lowell, Leominster, northeastern region